The techniques of deep freezing have already caused a revolution in the kitchen. The owner of the deep freeze can plan her menus for several weeks or even months ahead, and need no longer fear the arrival of unexpected guests, or the chore of shopping several times a week.

This book is a sequel to Mary Norwak's earlier success, *A Complete Guide to Deep Freezing*, which provided the basic information on freezing, packaging and storing. *Deep Freezing Menus and Recipes* includes over 200 further recipes of all types of dishes from fish, meat and poultry to pastry, puddings, cakes and breads.

The author has also added a comprehensive table of storage times and a special chapter on how to pack cooked dishes.

*Also by Mary Norwak
and available in Sphere Books*

A COMPLETE GUIDE TO
DEEP FREEZING

DEEP FREEZING MENUS AND RECIPES

Deep Freezing Menus and Recipes

MARY NORWAK

SPHERE BOOKS
30/32 Gray's Inn Road, London, WC1X 8JL

First published in Great Britain in 1970 by Sphere Books
Reprinted March, 1971
Reprinted January 1972
Copyright © Mary Norwak 1970

TRADE MARK

SPHERE

*Conditions of Sale – This book shall not without the
written consent of the Publishers first given be lent,
re-sold, hired out or otherwise disposed of by way of
trade in any form of binding or cover other than that in
which it is published.*

*This book is published at a net price and is
supplied subject to the Publishers Association
Standard Conditions of Sale registered under the
Restrictive Trade Practices Act 1956*

Set in Monotype Plantin

Printed in Great Britain by
C. Nicholls & Company Ltd.

RECIPES FOR THE FREEZER

CONTENTS

Introduction 9

CHAPTER
1. Table of Storage Times 11
2. How to Prepare and Pack Cooked Dishes 13
3. Starters 22
4. Main Courses 31
5. Sweet Courses 41
6. Pies, Pastry and Pie Fillings 48
7. Pasta 55
8. Ices and Frozen Puddings 60
9. Cakes and Biscuits, Icings and Fillings 67
10. Yeast Breads and Buns, Scones and Pancakes 77
11. Sandwiches and Fillings 87
12. School Holiday Food 90
13. Food for One 96
14. Picnics 98
15. Dinner Parties 103
16. Preparing for Christmas 107
17. Party Pieces 112
18. Food for Babies and Small Children 118
19. Preserves from the Freezer 121
20. Complete Meals from the Freezer 127
21. How to Use Frozen Food 130
22. Leftovers 136

Index 142

INTRODUCTION

Novice freezer owners tend to think only in terms of preserving raw materials such as home-grown fruit and vegetables, or fresh poultry and game. Experience brings the knowledge that frozen *prepared* dishes are invaluable to the busy housewife or career woman, to those who entertain a lot, and even to those who live alone. It is an economy in both shopping and cooking time to prepare dishes in bulk for future use. With the aid of today's mixers, two cakes may be prepared in the time formerly taken for one, so that not only is a fresh cake available on baking day, but a spare one is ready to be served, and just as fresh, a couple of months later. Likewise, a double quantity of any casserole can be halved, so that the same dish can be served at a later date.

Planning for school holidays or festivals such as Christmas becomes easier when whole meals and traditional dishes are stored away weeks ahead when more time is available for shopping and cooking. Unexpected guests or impromptu parties are never a problem when there is a supply of ready-cooked food on hand.

Experienced cooks and freezer owners are able to adapt their favourite recipes for freezer use, and work out their own methods of packing, thawing and reheating, but even they can find that results are variable, or that their friends have personal tricks for preparing and improving certain items.

With this in mind, the recipes in this book have been specially prepared and tested to find what will give the best results under the normal conditions which prevail in a busy housewife's kitchen. Packing materials and storage times are given which have been found most satisfactory, though it is often possible to store food longer than the stated times without dire results. Cooked food in the freezer however should be used up reasonably quickly and a rapid turnover maintained, or else the freezer becomes just another store cupboard, instead of a valuable time-saver from which cooked dishes can be chosen to complement the fresh seasonal food which is available.

The experienced freezer owner who caters for a family and for many guests tends to think in terms of meals for occasions rather than individual raw materials or dishes. For this reason, many chapters contain recipes for main courses, vegetables sauces and puddings in the groups in which they are most likely to be served, but individual recipes may be quickly located through the index.

Chapter One

TABLE OF STORAGE TIMES

It is difficult to give a firm rule for determining storage times for food kept in the freezer. The quality of the food served will depend on its initial quality before preparation, the rate of freezing, the suitability of packaging material, the efficiency of sealing, variations in temperature in the freezer, and the correct thawing and reheating methods.

Bad packing, leaving air spaces and the wrong packaging material will all affect the ultimate quality, flavour and texture of cooked dishes. Salt will shorten storage life, excess fat will cause rancidity, and some seasonings may develop off-flavours which may be unpleasant. Attention to hygiene and to thorough cooling before freezing are particularly important when dealing with cooked dishes.

This chart gives reasonable maximum safe storage times, but indivdiual recipes indicate specific details of storage and thawing times. Aim at a short storage period for cooked foods, with a rapid turnover.

Item	*Maximum Safe Storage (Months)*
Soups and Sauces	2 months
Cooked Fish Dishes	1 month
Cooked Meat Dishes	1 month
Meat Pies	2 months
Sandwiches	3 months
Yeast Bread, Rolls and Buns	
Baked	8–12 months
Unbaked	2 months
Biscuits	
Baked and Unbaked	6 months

Cakes
    Baked   3–4 months
    Unbaked   1 month
Pies
    Baked   6 months
    Unbaked   1–2 months
Steamed and Baked Puddings   3 months
Mousses, Soufflés, etc.   1 month

Chapter Two

HOW TO PREPARE AND PACK COOKED FOODS

To achieve perfect results, it is most important that cooked foods are correctly packaged and sealed for the freezer. Packaging need not be expensive, since many everyday household materials can be used. If incorrectly wrapped, food will dry out, and the presence of air will cause deterioration and the crossing of smells and flavours.

Packaging must be easy to handle, and must withstand low freezer temperatures, and should not be liable to splitting, bursting or leakage. Packaging materials should be moisture- vapour-proof, waterproof, greaseproof and smell-free.

BASIC PACKAGING MATERIALS

Waxed Tubs are suitable for soups, purées, casseroles and other foods which contain a quantity of liquid. Lids should be flush and airtight (spare lids are available for re-use). Screw-top tubs are also available.

Waxed Cartons with Fitted Lids are usually rectangular, and are useful for packing carved cooked meats without sauce or gravy, and for cakes and biscuits to avoid crushing.

Waxed Boxes with Tuck-in Lids are tall containers which are useful in limited space, and are available in ½ pint, 1 pint and 2 pint sizes. They are very good for soups, purees and fruit cooked in syrup.

Waxed Cartons with Liners are usually made in 2 pint size with a polythene liner which is useful for foods particularly subject to leakage.

Rigid Plastic Boxes with close-fitting lids are economical in use as they can be re-packed almost indefinitely. They are useful for sandwiches or other items for packed meals as they can be taken straight from the freezer and transported without repacking. They are also useful for stews which can be turned into a saucepan or casserole for thawing, since the flexible sides can be lightly pressed to aid removal. There are special Swedish freezer boxes which can be boiled for sterilization and which stack and save space, but efficient boxes can be bought at chain stores and ironmongers. These plastic boxes are often used for packing butter and other groceries and are worth saving to use in the freezer.

Glass Jars can be used for freezing if tested first for resistance to low temperatures. Screw top preserving jars, bottles and honey jars can be used, but is better to avoid jars with "shoulders" as the contents cannot then be removed quickly for thawing. To test jars, put an empty jar into a plastic bag and into the freezer overnight. If the jar breaks, the bag will hold the pieces. Oven glass can be used for the freezer, though this is an expensive way of packing, since the utensils are then out of use in the kitchen; they cannot be taken straight from the freezer to oven unless they are of the special type advertised for this purpose.

Polythene Bags of a heavy quality suitable for freezing are useful for holding bread, cakes, pies and sandwiches, and for collecting together items wrapped in foil such as soup cubes.

Foil is invaluable for preparing and freezing cooked food. Pie and pudding dishes and patty pans are useful for all types of pastry, and for many puddings and savoury dishes, and may be used straight from freezer to oven for heating. Compartment trays are also available for preparing and freezing whole meals. The food is normally cooked in the foil dishes, then packed for freezing in polythene bags, or in heavy duty foil. Heavy duty foil is an excellent wrapping for cakes and solid cooked foods such as patés. It is useful for making lids for foil or oven glass containers, and foil wrapped items can be put straight into the oven for reheating.

Fastening Materials are essential for successful freezing. Special freezer sealing tape is made with gum which is resistant to low temperatures so that it will not loosen and curl, and should be used for waxed and rigid plastic containers and for foil wrappings to seal all lids and seams. Polythene bags may be sealed with a special unit, but can be successfully closed with bag fasteners.

Labelling Materials should not be ignored. Wax or felt pencils must be used to avoid fading. Labels may be inserted in transparent packages, but if a label is put outside a package it must have special freezer-proof gum.

PACKING FOR THE FREEZER

Sheet Wrapping for baked goods and solid cooked foods should be finished like an old-fashioned chemist's parcel. Food should be placed in the centre of the heavy duty foil, then two sides of the sheet drawn together and folded neatly downwards over the food. This fold should be sealed, then the ends folded like a parcel, close and tight to the food to exclude air, and all folds sealed.

Bag Wrapping is not suitable for very liquid food. The bag should be completely open before filling, and a funnel used for any liquid to avoid messiness. Food must go down into the corners of the bag and leave no air pockets. Air must be extracted from both liquid foods and from items like cakes packed in bags, and this is most easily done by inserting a drinking straw and sucking out air from the bag with the neck held closely to the straw. Seal bags with heat or a tie fastener. To make bags easier to handle and store, they may be placed in rigid containers for filling and freezing, then removed in a more compact form.

Cartons and Rigid Containers which are used for liquid foods should not be completely filled to allow for expansion of the contents. The headspace between the surface of the food and the seal of the container is usually ½ ins. but in larger containers up to 1 ins. headspace is necessary.

Labelling is particularly important for cooked foods which are often unidentifiable by appearance alone. They are best labelled with the names of the contents, number of portions and date of freezing. Extra information which is useful includes special thawing, heating or seasoning instructions.

Freezing temperatures are important for cooked dishes, and they should be normally frozen in the coldest part of the freezer if a fast freezing adjustment cannot be made. Cooked food should be frozen in contact with the sides or bottom of the freezer. Manufacturer's instructions will indicate any special temperature control, the amount of food to be frozen at any one time, and the length of time a packet should be frozen before moving

A. TUBS: Made to stack on top of one another. For soups, purées, casseroles.

B. CARTONS: With fitted lids. For carved meats (without liquid) cakes, biscuits.

D. BAG FASTENERS: Gives air tight closure.

C. BOXES: With tuck in lids. Good shape for tucking in corners. (½, 1, 2 pint sizes)

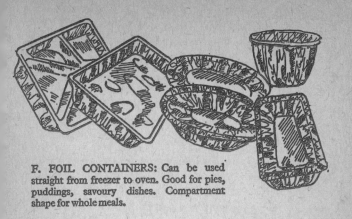

F. FOIL CONTAINERS: Can be used straight from freezer to oven. Good for pies, puddings, savoury dishes. Compartment shape for whole meals.

E. SWEDISH BOXES: Plastic. Stackable. Can be boiled. Good for sandwiches, biscuits.

G. HONEY, COFFEE, PRESERVING JARS: See Chap. 2 for testing before use.

H. OVEN GLASS: Good for using straight from freezer to oven to table but uneconomical in use.

to storage position. Bread and cakes do not however need fast freezing.

Storing cooked food for any length of time is unwise, and a quick turnover of cooked dishes should be aimed at. It is a good idea to keep cooked dishes such as casseroles, or baked goods, in one section of the freezer so that the amount available can be quickly seen. If baskets are not available or not practicable because of space, similar items may be grouped together in a nylon shopping bag.

Keeping Records is particularly important when storing cooked food which is normally used more quickly than raw materials. This can be kept in an exercise book, on a chart, or in its simplest form on a plastic shopping list which can be wiped when alterations are made. Useful information to be indicated includes position in the freezer, the type of item, the size of pack in weight of portions, the number available, the date on which item was frozen and the date by which it should be used.

AVOIDING DISASTER

Frozen food must not only be delicious to taste, but it must also look good and contain some nutritive value. Mistakes in freezing are costly in terms of not only raw materials and packaging, but also in time spent on cooking.

It is extremely important that cooked food should be prepared under hygienic conditions, cooled completely, packaged in the correct wrappings, sealed completely, and frozen quickly. To complete the picture, it is important that food should be thawed and reheated or cooked properly, so that the result is indistinguishable from freshly cooked food.

Food is spoiled by enzymic action and the aim in freezing is to retard this process. An enzyme is a type of protein which speeds up chemical reactions and causes harmful bacteria and cooked food is particularly subject to such action. When food is thawed, the enzymic action is speeded up, and deterioration is accelerated, and it is important that food is thawed and heated quickly afer freezing. *It is above all of utmost importance that no cooked dish should be thawed and then frozen again.*

Dehydration occurs when moisture and juices leave the food, and this may occur after a long period of storage; it is avoided by careful wrapping in the correct materials and the exclusion of air.

Oxidation occurs when air moves inwards to the food and the oxygen mingles with fat cells, reacting to form chemicals which give meat and fish a bad taste and smell; it is avoided by using the correct wrappings to provide an oxygen barrier.

Freezer Burn is the effect of dehydration which causes discoloured greyish-brown areas on the surfaces of food; it is avoided by using the correct packaging materials, wrapping to exclude air and sealing.

Rancidity is the effect of oxidation, and causes an unpleasant flavour and smell. Fatty foods suffer from this problem, and excess fat should be removed from cooked dishes before freezing. Salt can also accelerate the reaction which causes rancidity.

Cross-Flavouring occurs when strongly-flavoured foods affect other items in storage. Onions, garlic and some spices cause trouble, and cooked dishes containing these should be overwrapped.

Ice Crystals may form on such liquid foods as soup if too great headspace is left in the container. This is not a serious fault in liquids which will be heated or thawed, as the liquid melts back into its original form and can be shaken or stirred back into emulsion.

WHAT TO COOK FOR FREEZING

Before cooking foods for freezing, it is best to assess those items which are worth freezer space. Briefly, these are:

Dishes which need long cooking or long and tedious preparation.

Dishes made from seasonal foods.

Dishes which can be made in large quantities with little more work (i.e. three cakes instead of one; double or treble casseroles).

Dishes for special occasions, such as parties or holidays.

Convenience foods for invalids, small children, unexpected illness.

WHAT TO AVOID COOKING FOR THE FREEZER

Very few items cannot be frozen successfully. Items to avoid are hard-boiled eggs, sour and single cream (i.e. less than 40% butterfat), custards, soft meringue toppings, mayonnaise and salad dressings, milk puddings.

Sauces thickened with flour tend to separate, and cornflour should be used instead.

Rice, spaghetti and potatoes become mushy in liquid and should be added during reheating to soups and stews.

Onion, garlic, spices and herbs sometimes develop a musty flavour, and quantities should be reduced in such dishes as casseroles, or the items should be added during reheating.

OVERCOOKING

Meat stews may be overcooked during the reheating. It is wise to reduce initial cooking time in normal recipes by about 30 minutes which allows for reheating. Fish also suffers from overcooking, and times should be adjusted accordingly. Vegetables in pies or stews may become flabby and are best added to dishes about 15 minutes before cooking is completed and the dish frozen, or added later during the reheating.

PREPARING COOKED FOOD FOR FREEZING

Strict hygiene must be observed in preparing cooked food for the freezer. Use fresh, good quality foods for cooking. Cool cooked food promptly and quickly by standing the container in cold water and ice cubes. If a leftover portion of a dish is to be frozen, heat it thoroughly and cool at once before freezing.

Fats can cause trouble in the freezer. Avoid the use of dripping unless it has been recently clarified. Remove surplus fat from food before chilling in the freezer. Fried foods should be well drained on absorbent paper before freezing, and must be very cold before packing to avoid sogginess.

Freezing before Packaging is recommended when the surface of the food or decoration might be spoiled by packaging material. Such foods include unbaked pies, piped potatoes, decorated cakes and puddings. Biscuits and small cakes may also be frozen unwrapped on metal sheets, then packed in boxes to avoid crushing.

THAWING AND SERVING

Instructions for thawing and/or heating and serving are given in individual recipes. The time taken for thawing may vary according to size, shape and textures of foods. Most foods are best thawed in a refrigerator, and will take some time and so are best put in the refrigerator overnight. Foods thawed either in the refrigerator or at room temperature should be kept in their freezer wrappings, which may be loosened in the case of cakes and puddings. Foil slows thawing rate and may be replaced with another wrapping. Unbaked pies may be transferred direct to a preheated oven, and an air vent cut for the escape of steam when the pie begins to heat through. Casseroles are best placed in a cold oven which should then be set to the temperature required and this will avoid scorching at the edges of dishes. Thawing before reheating is always best for dishes in which eggs or cream are used to thicken, and these are best reheated in a double boiler. Partial thawing is sometimes necessary to get food out of packaging.

Chapter Three

STARTERS

It is extremely useful to freeze a variety of first courses which can be used to extend a simple meal, or which will give a hostess time to concentrate on a freshly-cooked and perhaps elaborate main course. Additionally most of these items, such as soup and paté are excellent mainstays for luncheon and supper, the meal perhaps being completed with bread, butter, and fresh fruit.

SOUP

In addition to complete soups, meat, chicken and fish stock may all be frozen to use as a basis for fresh soups. These stocks should be strained, cooled and defatted, and packed into cartons with headspace. They are best thawed in a saucepan over low heat.

Soup which is thickened with ordinary flour tends to curdle on reheating, and cornflour is best used as a thickening agent, and gives a creamy result. Rice flour can be used, but makes the soup glutinous. Porridge oats can be used for thicker meat soups. Starchy foods such as rice, pasta, barley and potatoes become slushy when frozen in liquid, and should be added during the final reheating after freezing. It is also better to omit milk or cream from frozen soups, as results with these ingredients are variable, and they can be added when reheating.

Soup to be frozen should be cooled and surplus fat removed as this will separate in storage and may cause off-flavours. Soup should be frozen in leakproof containers, allowing ½ ins. headspace for wide-topped containers and ¾ ins. headspace for narrow-topped containers. Rigid plastic containers are useful for storage, but large quantities of soup may be frozen in bread tins or freezer boxes lined with foil; the solid block can then be wrapped in foil and stored like a brick.

Soup should not be stored longer than 2 months. It will thicken during freezing, and allowance should be made for this in the recipe so that additional liquid can be added on reheating without spoiling the soup. Seasonings may cause off-flavours, and it is best to season after thawing. Clear soups can be heated in a saucepan over low heat, but cream soups should be heated in a double boiler and well-beaten to keep them smooth.

SOUP GARNISHES

Herbs and croutons can be frozen to give an attractive finish to soups even when time is limited. *Herbs* such as parsley and chives should be chopped and packed in ice cube trays with a little water, then each frozen cube wrapped in foil. The herb cubes can be reheated in the soup. *Croutons* can be prepared from lightly toasted ½ ins. slices of bread which are then cut in cubes and dried out in an oven set at 350°F (Gas Mark 4). They are best packed in small polythene bags and thawed in wrappings at room temperature or reheated if preferred. As a variation, the bread can be toasted on one side only and the other side spread with grated cheese mixed with a little melted butter, egg yolk and seasoning, which is then toasted and the bread cut in cubes before packing.

SAVOURY FLANS

Baked pastry flan cases can be frozen and quickly reheated and filled for first courses or snacks (see PIES, PASTRY AND PIE FILLINGS). A few fillings may be frozen in the flans to save time when they are required to be eaten.

PÂTÉ

Pâté freezes extremely well. It can be packed in individual pots for serving, or cooked in loaf tins or terrines, then turned out and wrapped in foil for easy storage.

SAVOURY PANCAKES

Pancakes may be frozen for use with savoury or sweet fillings (see YEAST BREADS AND BUNS, TEA BREADS, SCONES, PANCAKES). They may also be filled and frozen ready for use, with or without a suitable sauce. Suitable fillings include creamed chicken and mushrooms; prawns or lobster in a cream sauce; sweetbreads, chicken livers or ham in wine sauce; creamed smoked haddock; spinach and cheese. The filled pancakes can be covered with a mushroom, cheese or wine sauce before freezing; or finished with a sauce or grated cheese only at the reheating stage.

Further ideas for first courses are included in the menus for DINNER PARTIES, and in the chapter on PASTA DISHES.

TOMATO SOUP

2 lbs. tomatoes	3 pints stock
2 oz. mushrooms	2 oz. rice flour
2 medium onions	2 egg yolks
1 leek	¼ pint creamy milk
2 sticks celery	Pinch of sugar
Juice of 1 lemon	Salt and pepper
Parsley, thyme, bayleaf	Red colouring
2 oz. butter	

Cut tomatoes in slices. Slice mushrooms, onions, leek and celery and cook lightly in butter. Add the lemon juice, herbs and stock and tomatoes and simmer for 30 minutes. Sieve the mixture. In a bowl, mix egg yolks, rice flour and milk until creamy and add a little of the hot tomato mixture, stirring gently. Add remaining liquid and cook very gently for 10 minutes without boiling. Season to taste with salt and pepper and sugar and colour if necessary.

Pack after cooling into cartons, leaving headspace.
To serve reheat in a double boiler, stirring gently.
Storage Time 2 months.

BROWN VEGETABLE SOUP

2 carrots	Parsley, thyme, bayleaf
2 turnips	Salt and pepper
2 onions	1 oz. cornflour
1 quart beef stock	

Roughly cut carrots and turnips into pieces. Slice onions and brown in a little butter. Add stock and other vegetables with herbs and seasoning, and simmer for 1 hour. Sieve and thicken with cornflour, and simmer for 5 minutes.

Pack after cooling into cartons, leaving headspace.
To serve reheat in a double boiler, stirring gently.
Storage Time 2 months

SCOTCH BROTH

1 lb. lean neck of mutton	1 carrot
4 pints water	1 turnip
1 leek	Sprig of parsley
2 sticks celery	Salt and pepper
1 onion	

Cut meat into small squares and simmer in water for 1 hour. Add vegetables cut in dice, parsley and seasoning, and continue cooking gently for 1½ hours. Cool and remove fat and take out parsley.

Pack into containers, leaving headspace.

To serve reheat gently in saucepan and add 2 tablespoons barley, simmering until barley is tender.

Storage Time 2 months.

OXTAIL SOUP

1 oxtail	1 turnip
2½ pints water	1 stick celery
2 carrots	Salt
2 onions	

Wipe oxtail and cut in pieces. Toss in a little seasoned flour and fry in a little butter for 10 minutes. Put in pan with water and simmer for 2 hours. Remove meat from bones and return to stock with vegetables cut in neat pieces. Simmer for 45 minutes and put through a sieve, or liquidise. Cool and remove fat.

Pack in containers, leaving headspace.

To serve reheat gently in saucepan, adding ½ teaspoon Worcestershire sauce and ½ teaspoon lemon juice.

Storage Time 2 months.

KIDNEY SOUP

8 oz. ox kidney	1 carrot
1 oz. butter	Parsley, thyme, bayleaf
1 small onion	Salt and pepper
1 quart stock	1 oz. cornflour

Wash kidney and cut in slices. Cook kidney and sliced onion in butter until onion is soft and golden. Add stock, chopped carrots, herbs and seasoning and simmer for 1½ hours. Put through

a sieve, or liquidise, and return to pan. Thicken with cornflour and simmer for 5 minutes.
Pack after cooling into containers, leaving headspace.
To serve reheat in double boiler, stirring gently and adding ½ gill sherry.
Storage Time 2 months

ONION SOUP

1½ lbs. onions	Salt and pepper
2 oz. butter	2 tablespoons cornflour
3 pints beef stock	

Slice the onions finely and cook gently in butter until soft and golden. Add stock and seasoning, bring to the boil, and simmer for 20 minutes. Thicken with cornflour and simmer for 5 minutes.
Pack after cooling and removing fat in containers, leaving headspace.
To serve reheat in double boiler, stirring gently. Meanwhile, spread slices of French bread with butter and grated cheese, and toast until cheese has melted. Put slices into tureen or individual bowls and pour over soup.
Storage Time 2 months.

SHRIMP BISQUE

2 sticks celery	Salt and pepper
4 oz. mushrooms	Bayleaf
1 small onion	Pinch of nutmeg
1 carrot	2 tablespoons lemon juice
2 oz. butter	2 tablespoons white wine
2 pints chicken stock	6 oz. shrimps

Cut celery, mushrooms, onion and carrot in small pieces and cook gently in butter for 10 minutes. Add stock, seasoning, bayleaf, nutmeg and lemon juice, and simmer for 20 minutes. Put through a sieve. Add wine and shrimps and simmer for 5 minutes. Cool and remove fat.
Pack into containers, leaving headspace.
To serve reheat in a double boiler, stirring gently. When thawed,

stir in ½ pint double cream and continue reheating without boiling.
Storage Time 1 month.

CHICKEN LIVER PÂTÉ

8 oz. chicken livers	1 small onion
3 oz. fat bacon	1 egg
2 crushed garlic cloves	Salt and pepper

Cut livers in small pieces and cut up bacon and onion. Cook bacon and onion in a little butter until onion is just soft. Add livers and cook gently for 10 minutes. Mince very finely and season. Add garlic and beaten egg, and put mixture into foil containers. Stand containers in a baking tin of water and cook at 350°F (Gas Mark 4) for 1 hour. Cool completely.
Pack by covering each container with a lid of foil, sealing with freezer tape.
To serve thaw at room temperature for 1 hour. Use immediately after thawing.
Storage Time 1 month.

PHEASANT PÂTÉ

8 oz. calves liver	Salt and pepper
4 oz. bacon	cooked pheasant
1 small onion	Powdered cloves and allspice

Cook liver and bacon lightly in a little butter, and put through mincer with onion. Season with salt and pepper. Remove meat from pheasant in neat pieces and season lightly with cloves and allspice. Put a layer of liver mixture into dish (foil piedish, loaf tin, or terrine), and add a layer of pheasant. Continue in layers finishing with liver mixture. Cover and steam for 2 hours. Cool with heavy weights on top.
Pack by covering container with foil lid and sealing with freezer tape; or by turning pate out of cooking utensil and wrapping in heavy duty foil.
To serve thaw in wrappings in refrigerator for 6 hours, or at room temperature for 3 hours.
Storage Time 1 month

HARE PÂTÉ

1½ lbs. uncooked hare	¾ lb. minced pork and veal
¼ lb. fat bacon	Salt, pepper and nutmeg
3 tablespoons brandy	1 egg

This recipe may also be used for a mixture of game or for rabbit. Cut hare into small pieces and bacon into dice and mix together in a dish with brandy. Leave for 1 hour, then put through mincer with pork and veal. Season, add egg and mix well. Press mixture into a buttered container, cover with greased paper and lid, and put dish in a baking tin of water. Bake at 400°F (Gas Mark 6) for 1 hour. Leave under weights until cold.

Pack by covering container with foil lid and freezing tape, or by repacking in sheet heavy duty foil. This is only advisable if a large amount of pâté is to be eaten at once. Otherwise, repack mixture into small containers, cover and freeze (if the pâté is cooked in small containers, it will be dry).

To serve thaw small containers at room temperature for 1 hour. Thaw large pâté in wrappings in refrigerator for 6 hours, or at room temperature for 3 hours. Use immediately after thawing.

Storage Time 1 month.

SIMPLE PORK PÂTÉ

¾ lb. pig's liver	1 tablespoon flour
2 lbs. belly of pork	Salt, pepper and nutmeg
1 large onion	Parsley
1 large egg	Streaky bacon

Put liver and pork through coarse mincer. Chop onion and soften in a little butter. Mix together meat, onion, egg beaten with flour, seasoning and a little chopped parsley. Line foil dish or terrine or loaf tin with rashers of streaky bacon flattened with a knife. Put in mixture. Cover with greaseproof paper and lid, and stand container in a baking tin of water. Cook at 350°F (Gas Mark 4) for 1½ hours. Cool under weights.

Pack by covering container with foil lid and sealing with freezer tape, or remove from container and wrap in heavy duty foil.

To serve thaw in wrappings in refrigerator for 6 hours or at room temperature for 3 hours.

Storage Time 2 months.

COD'S ROE PÂTÉ

12 oz. smoked cod's roe
1 gill double cream
1 crushed garlic clove
Juice of ½ lemon
1 dessertspoon olive oil
Black pepper

Scrape roe into bowl and mix to a smooth paste with cream, garlic, lemon, oil and pepper.
Pack into small containers with lids.
To serve thaw in refrigerator for 3 hours, stirring occasionally to blend ingredients.
Storage Time 1 month.

POTTED SHRIMPS

Shrimps
Butter
Salt and pepper
Ground mace and cloves

Cook freshly caught shrimps, cool in cooking liquid and shell. Pack tightly into waxed cartons. Melt butter, season with salt, pepper, and a little mace and cloves. Cool butter and pour over shrimps. Chill until cold.
Pack by covering with lids and sealing with freezer tape.
To serve thaw in containers at room temperature for 2 hours, or heat in double boiler until butter has melted and shrimps are warm to serve on toast.
Storage Time 6 months.

QUICHE LORRAINE

4 oz. short pastry
½ oz. butter
1 small onion
2 oz. streaky bacon
1 egg and 1 egg yolk
2 oz. grated cheese
Pepper
1 gill creamy milk

Line a flan ring with pastry or line foil dish which can be put into freezer. Gently soften chopped onion and bacon in butter until golden, and put into pastry case. Lightly beat together egg, egg yolk, cheese, pepper and milk (add a little salt if the bacon is not very salt.) Pour into flan case. Bake at 375° F (Gas Mark 5) for 30 minutes. Cool.
Pack in foil dish in rigid container to avoid breakage, and seal with freezer tape.

To serve thaw in refrigerator for 6 hours to serve cold. If preferred hot, heat at 350°F (Gas Mark 4) for 20 minutes.
Storage Time 2 months.

SAVOURY PANCAKES

4 oz. plain flour	1 egg and 1 egg yolk
Pinch of salt	½ pint milk
1 tablespoon melted butter	

Make batter and cook very thin pancakes. These may be large, or about 5 ins. in diameter. Spread pancakes with desired filling and fold in half twice to form quarter-circles.
Pack in suitable quantities, separating each with a sheet of Cellophane. Or arrange pancakes in foil tray and cover with sauce.
To serve arrange pancakes in dish, cover with sauce, or grated cheese and reheat in moderate oven. Pancakes already frozen in sauce can also be reheated in moderate oven.
Storage Time 1 month.

Chapter Four

MAIN COURSES

Meat and poultry can be usefully frozen in the form of main dishes. These may be specially prepared in large quantities, or the normal family-size recipe doubled and divided in half, part for immediate use and the second half for freezing, and this is the most useful method for busy housewives.

There is little advantage in pre-cooking joints, steaks, chops, or whole birds, since the outer surface sometimes develops an off-flavour, reheating it will dry out the meat, and if the meat is eaten without reheating it will tend to flabbiness. Cold meat or poultry may be frozen in slices, with or without sauce or gravy. Fried meats are rarely successful, since they tend to toughness, dryness and rancidity when frozen. Combination dishes of meat and vegetables should include the vegetables when they are slightly undercooked to avoid softness on reheating. *All cooked meat dishes must be cooled quickly before freezing.*

Roast and fried poultry are not worth freezing to eat cold, because on thawing they exude moisture and become flabby. Boiling chickens are worth preparing specially for the freezer; the meat being frozen ready-sliced or minced, or made at once into pies and casseroles for freezing, and the cooking liquid reduced to strong stock for freezing.

CASSEROLES AND STEWS

There must be plenty of liquid in frozen casseroles and stews to cover the meat completely, or the meat may dry out. Vegetables are best slightly undercooked to avoid softness. Potatoes, rice or other starch additions should be made during thawing and heating for service, as they become soft when frozen in liquid and often develop off-flavours. Sauces and gravies tend to thicken during storage; ordinary flour in a recipe may result in curdling during reheating, and cornflour should be substituted. While almost any recipe can be adapted for freezer use, the fat content should be as low as possible to avoid rancidity. When a dish is cooled before freezing, surplus fat should be removed from the surface.

GALANTINES AND MEAT LOAVES

Galantines are most easily used if cooked before freezing ready to serve cold. Meat loaves may be frozen uncooked, and this is made easy if the mixture is packed into loaf tins lined with foil, the foil then formed into a parcel for freezing; the frozen meat loaf can then be returned to the original tin for baking. Galantines can be prepared directly in loaf tins, then turned out, wrapped and frozen.

For cold serving, these compact meats may be packed in slices, divided by Cellophane or greaseproof paper, and re-formed into a loaf shape for freezing. Slices can be separated while still frozen and thawed quickly on absorbent paper.

SLICED MEAT AND POULTRY

Cold meat and poultry may be frozen in slices to serve cold. Slices should be at least ¼ ins. thick, separated by Cellophane or greaseproof paper, and must be packed tightly to avoid dry surfaces, then put into cartons or bags. They should be thawed for 3 hours in a refrigerator in the container, then separated and placed on absorbent paper to remove moisture. Ham and pork will lose colour when frozen like this.

It is preferable to freeze meat and poultry slices in gravy or sauce to retain juiciness. The liquid should be thickened with cornflour, and both meat and gravy or sauce should be cooled quickly before packing. These slices are best packaged in foil containers, covered with a lid, and this can save time in reheating as the container can go straight into the oven and the meat will remain moist. The frozen slices in gravy should be heated for 30 minutes at 350°F (Gas Mark 4).

PREPARING AND COOKING COOKED DISHES FOR FREEZING

It is most important that strict hygienic conditions be observed when preparing cooked meat or poultry dishes for the freezer. Leftovers should not be left in a warm kitchen, or stored even in a refrigerator before conversion into cooked dishes, and fresh raw materials should also be used as quickly as possible. Dishes must be completely cold before freezing, and the cooling process must be carried out quickly, most easily by standing the container of food in a bowl of iced water. Where ingredients such as meat and gravy are to be combined they should be thoroughly chilled before mixing, and for instance hot gravy should not be poured over cold meat.

PACKING AND SERVING

The methods of packing galantines, meat loaves and sliced meat are described above. When to be eaten cold, they should be thawed in wrappings in the refrigerator for 3 hours before serving. To be served hot, they can be transferred still frozen into the oven to heat for the time specified in the recipe.

Casseroles and stews are best frozen in foil containers which can be used in the oven, or in foil-lined containers so that the foil can be formed into a parcel for freezing, and the contents returned to the original container for heating and serving. If frozen in cartons, the dishes can be transferred to ovenware, or reheated in a double boiler or over direct heat if curdling is not likely to occur. It is expensive to freeze in ovenware dishes, though this is possible, as the dishes are then out of normal use in the kitchen; also they must be returned to room temperature before putting into the oven unless they are of the special freezer-to-oven variety, and this is of course, time-wasting.

STORAGE LIFE

Cooked meat dishes are best frozen for no longer than 1 month. Herbs and spices may develop off-flavours, fat become rancid, or vegetables flabby under longer storage. In practice, many dishes can be stored longer, but it is better freezer practice to maintain a frequent turnover of dishes, and plenty of variety.

BEEF IN WINE

3 lbs. shin beef	2 oz. bacon
1½ oz. butter.	Thyme and parsley
1½ oz. oil	1 tablespoon tomato purée
1 medium onion	Stock
2 garlic cloves	½ pint red wine

Cut meat into slices and cover very lightly with seasoned flour. Fry in a mixture of butter and oil until meat is just coloured, then add sliced onion, crushed garlic and bacon cut in small strips. Add herbs and wine and cook quickly until liquid is reduced to half. Work in tomato purée and just cover in stock, then simmer for 2 hours. Remove herbs and cool.

Pack into waxed or rigid plastic containers, or into foil-lined dish, forming the foil into a parcel, and removing for storage when frozen.

To serve reheat in double boiler.

Storage Time 1 month.

JELLIED BEEF

4 lbs. beef brisket	½ pint stock
8 oz. lean bacon	Pinch of nutmeg
Salt and pepper	Parsley, thyme and bayleaf
1 pint red wine	4 onions
2 oz. butter	4 carrots
2 oz. oil	1 calf's foot

The meat should not have too much fat, and should be firmly tied. Put meat to soak in wine for 2 hours after seasoning well with salt and pepper. Drain meat and brown in butter and oil, then put into casserole with wine, stock, nutmeg, herbs, sliced onions and carrots, and split calf's foot. Cover and cook at 325°F (Gas Mark 3) for 3 hours. Cool slightly and slice beef. Put meat into containers with vegetables. Strain liquid, cool and pour over meat and vegetables.

Pack in waxed or rigid plastic containers, or in foil-lined dish, forming the foil into a parcel, and removing for storage when frozen.

To serve thaw in refrigerator for 3 hours to eat cold. To eat hot,

put in covered dish in moderate oven (350°F or Gas Mark 4) for 45 minutes.
Storage Time 1 month

LAMB CURRY

4 tablespoons oil	1 large cooking apple
3 lbs lamb shoulder cut into cubes	1 teaspoon salt
	1 bayleaf
1 clove garlic	2 teaspoons grated lemon peel
2 large onions	1 tablespoon soft brown sugar
2 tablespoons curry powder	2 tablespoons sultanas

Heat oil and brown lamb cubes on all sides. Remove meat from oil. Add crushed garlic, chopped onion, curry powder and chopped apple, and toss over heat for 5 minutes until onion is soft. Add ¼ pint water, mixing well. Add lamb cubes, salt, bayleaf, lemon peel, sugar and sultanas and simmer with a lid on for 1½ hours, until liquid is reduced and lamb is tender. Cool.
Pack in waxed or rigid plastic containers.
To serve heat in double boiler with lid on, stirring occasionally.
Storage Time 1 month.

VEAL IN TOMATO SAUCE

4 lbs. shoulder veal cut into cubes	Salt and pepper
	4 large tomatoes
3 tablespoons olive oil	½ pint dry white wine
1 garlic clove	¾ pint chicken stock
1 lb. mushrooms	Parsley, thyme and bayleaf
20 small white onions	

Brown the veal cubes in oil until golden. Remove veal from oil, and cook crushed garlic, sliced mushrooms, and small whole onions until just soft. Remove vegetables and mix with meat. Into the pan juices, stir the tomatoes which have been peeled, seeded and chopped, together with salt and pepper, wine, stock and herbs. Cook gently until sauce is smooth, thickening a little with cornflour if liked. Cool sauce and pour over meat and vegetables, removing herbs.
Pack in waxed or rigid plastic containers, or in foil-lined dish,

forming the foil into a parcel, and removing for storage when frozen.
To serve put into oven still frozen and cook at 325° F (Gas Mark 3) for 1½ hours.
Storage Time 1 month.

VEAL WITH CHEESE

1½ lbs. veal cut in thin slices	½ teaspoon sugar
4 tablespoons oil	Pinch of rosemary and thyme
2 medium onions	8 oz. Gruyere cheese
1 garlic clove	2 tablespoons Parmesan cheese
1 lb. can tomatoes	

Coat veal very lightly in seasoned flour and cook until lightly browned in oil. Remove veal from oil. Cook sliced onions and crushed garlic in oil until golden. Add sieved tomatoes, sugar and herbs and simmer for 5 minutes. Pour half the tomato sauce into casserole, top with veal and slices of Gruyere. Pour over remaining sauce and sprinkle with Parmesan. Bake at 350°F (Gas Mark 4) for 30 minutes. Cool.
Pack by covering container in foil (it is best if the casserole is made in dish in which it is to be served to avoid disturbing the topping).
To serve heat at 350°F (Gas Mark 4) without lid for 30 minutes.
Storage Time 1 month.

PORK WITH ORANGE SAUCE

3 lbs. lean pork chops	2 tablespoons vinegar
2 medium onions	1 tablespoon brown sugar
½ pint orange juice (fresh, tinned or frozen)	

Toss the meat very lightly in seasoned flour and cook in a little oil until browned. Remove from oil, and cook sliced onions until just soft. Return chops and onions to pan, pour over orange juice, vinegar and sugar and simmer gently for 30 minutes until chops are cooked through. Cool.
Pack in foil trays, covering with sauce, and with foil lid.
To serve heat with lid on at 350°F (Gas Mark 4) for 30 minutes. Garnish with fresh orange segments.
Storage Time 1 month.

KIDNEYS IN WINE

16 lamb's kidneys	½ pint red wine
8 oz. mushrooms	Salt and pepper
2 oz. butter	Cornflour

Prepare kidneys by cutting in half, removing skin, fat and tubes. Cook kidneys gently in butter until just coloured, but still soft. Add sliced mushrooms, wine and seasoning and simmer for 30 minutes. Thicken sauce with a little cornflour if liked. Cool.
Pack in waxed or rigid plastic containers.
To serve reheat in double boiler and garnish with chopped parsley.
Storage Time 1 month.

LIVER CASSEROLE

1 lb. calves' or lamb's liver	1 tablespoon chopped parsley
4 oz. breadcrumbs	Salt and pepper
2 large sliced onions	¾ pint stock

Lightly toss liver in seasoned flour, and brown lightly in a little butter. Put into casserole or foil dish (which can be used in the freezer) in layers with breadcrumbs, onions, parsley and seasonings, finishing with a layer of crumbs. Pour in stock. Cover with lid and bake at 350°F (Gas Mark 4) for 45 minutes. Cool.
Pack by covering dish with foil lid.
To serve remove foil lid and heat at 350°F (Gas Mark 4) for 45 minutes. Serve with additional gravy.
Storage Time 1 month.

MEAT BALLS

¾ lb. minced fresh beef	1 small chopped onion
¼ lb. minced fresh pork	1½ teaspoons salt
2 oz. dry white breadcrumbs	¼ teaspoon pepper
½ pint creamy milk	Butter

Mix together beef and pork and soak breadcrumbs in milk. Cook onion in a little butter until golden, and mix together with meat, breadcrumbs and seasonings until well blended. Shape into 1 ins. balls, using 2 tablespoons dipped in cold water. Fry balls in butter until evenly browned, shaking pan to keep balls round. Cook a few at a time, draining each batch, and cool.

Pack in bags, or in boxes with greaseproof paper between layers.
To serve thaw in wrappings in refrigerator for 3 hours and eat cold. To serve hot, fry quickly in hot fat, or heat in tomato sauce or gravy.
Storage Time 1 month.

OVEN-FRIED CHICKEN

2 lbs. chicken pieces	1 teaspoon Worcestershire
¼ pint sour cream	sauce
1 dessertspoon lemon juice	1 teaspoon salt
	Pinch of pepper and paprika

Fresh cream may be soured with 1 teaspoon lemon juice to ¼ pint cream, if commercial sour cream is not available. Canned evaporated milk may also be substituted successfully in this recipe. Mix together cream, lemon juice and seasonings, chopping garlic finely. Cover chicken pieces completely in mixture and coat in breadcrumbs. Arrange in baking dish which has been well greased. Bake at 350°F (Gas Mark 4) for 45 minutes. Cool completely.
Pack in waxed box or foil dish in a single layer, or wrap each chicken piece individually.
To serve bake complete chicken dish at 450°F (Gas Mark 8) for 45 minutes; individual chicken joints will take 30 minutes. Uncover during final 10 minutes to give crisp surface.
Storage Time 1 month.

CHICKEN IN CURRY SAUCE

3 lbs. chicken pieces	cooking chicken pieces)
2 medium onions	1 tablespoon vinegar
1 tablespoon curry powder	1 tablespoon brown sugar
1 tablespoon cornflour	1 tablespoon chutney
1 pint chicken stock (from	1 tablespoon sultanas

Simmer chicken in water until tender, drain off stock and keep chicken warm. Fry sliced onions in a little butter until soft, add curry powder and cook for 1 minute. Slowly add chicken stock and the cornflour blended with a little water. Add remaining ingredients and simmer for 5 minutes. Add chicken pieces and simmer 15 minutes. Cool.
Pack in waxed or rigid plastic containers.

To serve heat in double boiler.
Storage Time 1 month.

POT ROAST PIGEONS

4 pigeons	1 teaspoon mixed herbs
½ pint stock	

Clean and wipe pigeons and coat lightly with seasoned flour. Brown birds in a little butter, add stock and herbs and simmer very gently for 1 hour under a tight lid. Cool quickly.
Pack in a foil dish, cover with juices, and pack dish into polythene bag.
To serve put birds into tightly covered casserole and heat at 350°F (Gas Mark 4) for 1 hour; serve with pan juices, bread sauce and game chips.
Storage Time 1 month.

PIGEON CASSEROLE

2 pigeons	Salt and pepper
8 oz. chuck steak	1 tablespoon redcurrant jelly
2 rashers bacon	1 tablespoon lemon juice
½ pint stock	1 tablespoon cornflour
2 oz. small mushrooms	

Cut pigeons in halves and the steak in cubes, and cut bacon in small pieces. Cook pigeons, steak and bacon until just coloured in a little butter. Put into a casserole with stock, sliced mushrooms, salt and pepper and cook at 325°F (Gas Mark 3) for 1 hour. Stir in redcurrant jelly, lemon juice and cornflour blended with a little water, and continue cooking for 30 minutes. Cool.
Pack in foil-lined dish, forming foil into a parcel, and removing from dish when frozen, for easy storage.
To serve cook at 350°F (Gas Mark 4) for 1 hour.
Storage Time 1 month.

PHEASANT IN CIDER

1 old pheasant	1 garlic clove
1 lb. cooking apples	Bunch of mixed herbs
8 oz. onions	Salt and pepper
½ pint cider	

Clean and wipe pheasant. Cut apples in quarters after peeling and coring, and put into a casserole. Slice onions and cook until soft in a little butter. Put pheasant on to apples and cover with onions. Pour on cider and add crushed garlic and herbs and season with salt and pepper. Cover and cook at 325°F (Gas Mark 3) for 2 hours. Cool and remove herbs.

Pack in foil container, and strain sauce over pheasant. Cover with foil lid. Or pack in foil-lined casserole, removing after freezing for easy storage.

To serve put into casserole, and cook at 350°F (Gas Mark 4) for 1 hour.

Storage Time 1 month.

JUGGED HARE

1 hare	Salt and pepper
1 carrot	4 pints water
1 onion	2 oz. butter
1 blade mace	2 tablespoons oil
Parsley, thyme and bayleaf	1 tablespoon cornflour
4 cloves	½ pint port

Soak head, heart and liver of hare for 1 hour in cold salted water. Put into a pan with carrot, onion, mace, herbs, cloves, salt and pepper and water, and simmer for 3 hours, skimming frequently. Coat pieces of hare lightly in seasoned flour and brown in mixture of butter and oil. Put into a casserole. Strain stock and mix with cornflour blended with a little water. Simmer until reduced to 3 pints, and pour over hare. Cover and cook at 325°F (Gas Mark 3) for 4 hours. Remove hare pieces and cool. Add port to gravy and simmer until of coating consistency. Cool. Pack and cover with gravy.

Pack in waxed or rigid plastic containers, leaving ¾ ins. headspace for gravy.

To serve put into casserole and heat at 350°F (Gas Mark 4) for 45 minutes, adding fresh or frozen forcemeat balls 10 minutes before serving time.

Storage Time 1 month.

Chapter Five

SWEET COURSES

The sweet course of a meal is a constantly recurring problem for the busy housewife. Even though she may prefer to cook her main course from fresh raw materials each day, she can hardly fail to appreciate having a supply of frozen puddings ready to hand for both everyday and party occasions. Apart from the more obvious items such as ice cream and pastry which store so well in the freezer, frozen pancakes and spongecakes can quickly be converted with fruit, cream or sauces into delicious puddings.

Additionally, steamed puddings and crumbles can be frozen ahead for school holidays, gelatine sweets, cold soufflés and cheesecakes freeze well for weekends or parties, and fruit may be prepared in pudding form to save time and often freezer space. Only milk puddings are not really successful in the freezer, as they become mushy or curdled if containing eggs, and little time is saved by their advance preparation.

PUDDING AND CAKE MIXTURES

These may be made to standard recipes and either steamed or baked. They are most easily made in foil containers in which they can be frozen and reheated for serving. It is better not to put jam or syrup in the bottom of these puddings before cooking, but dried fruit, fresh fruit or nuts may be added. Highly-spiced puddings may develop off-flavours.

Suet puddings containing fresh fruit may be frozen raw or cooked. On balance, it is more useful to cook them before freezing, since only a short time need then be allowed for reheating before serving. Puddings made from cake mixtures or the traditional suet puddings can be made from any standard recipe, and may also be frozen raw or cooked; lemon, chocolate and dried fruit are particularly good flavourings for freezer storage. Cake mixtures may also be used to top such fruits as apples, plums, gooseberries and apricots; these are just as easily frozen raw since complete cooking time in the oven will be little longer than reheating time. This also applies to fruit puddings with a "crumble" topping.

GELATINE SWEETS

Many cold puddings such as jellies, mousses and soufflés involve the use of gelatine. When gelatine is used for creamy mixtures to be frozen, it is entirely successful, but clear jellies are not really recommended for the freezer. The ice crystals formed in freezing break up the structure of the jelly, and while it retains its setting quality, the jelly becomes granular and uneven and loses clarity. This granular effect is masked in such puddings as cold soufflés.

FRUIT SWEETS

Raw fruit can take up a lot of freezer space and also must be thawed and made into a dish before use. It is very convenient to prepare a few kinds of fruits in syrup, particularly good if flavoured with wine or liqueurs, which need no further cooking after freezing. These are particularly useful for using such fruits as pears and peaches which are difficult to freeze well in their raw state.

SAUCES

A supply of sweet sauces such as fruit sauce or chocolate sauce can be usefully frozen for use with puddings or ices. These are best prepared and frozen in small containers, and reheated in a double boiler.

STRAWBERRY MOUSSE

1½ lbs. frozen strawberries
4 teaspoons lemon juice
2 oz. sugar
1 oz. gelatine
1 pint double cream

Thaw strawberries, press through sieve with juice and add to lemon juice and sugar. Soften gelatine in 2 tablespoons cold water and stand in pan of hot water until gelatine is syrupy. Make up to ¼ pint with boiling water and stir into strawberry

mixture. Cool and then refrigerate for about 1 hour until the mixture is like unbeaten egg white. Whip cream until stiff and fold strawberry mixture into cream. Put into 7 ins. soufflé dish with a collar of paper or foil, and chill in refrigerator until firm. Put into freezer until solid and remove collar.
Pack in soufflé dish wrapped in foil.
To serve thaw in refrigerator for 12 hours and serve with cream.
Storage Time 1 month.

NESSELRODE MOUSSE

5 eggs
¼ teaspoon cream of tartar
4 oz. caster sugar
Pinch of salt
2 tablespoons light rum
2 teaspoons lemon juice
¾ pint double cream
4 oz. chopped mixed glacé fruit

Separate eggs and beat egg whites with cream of tartar until soft peaks are formed. Gradually add 3 oz. of the sugar beating well after each addition until peaks are stiff. Beat egg yolks and salt until thick and lemon coloured and gradually beat in remaining sugar. Continue beating while adding rum and lemon juice. Beat cream until stiff. Carefully fold rum mixture, cream and glacé fruit into egg white mixture and turn into serving bowl in which mousse is to be frozen.
Pack by wrapping serving dish in foil.
To serve leave at room temperature for 1 hour and garnish with cherries and angelica.
Storage Time 1 month

CHOCOLATE MOUSSE

8 oz. plain chocolate   4 eggs

Melt chocolate in double boiler (a little strong coffee may be added). Remove from heat and add egg yolks, one at a time, mixing well. Fold in stiffly beaten egg whites. Pour into individual dishes.
Pack by wrapping dishes in foil.
To serve thaw in refrigerator for 2 hours and serve with cream.
Storage Time 1 month.

LEMON PUDDINGS

2 oz. cornflakes	1 tablespoon grated lemon peel
3 eggs	3 tablespoons lemon juice
4 oz. caster sugar	½ pint double cream

Crush cornflakes and sprinkle a little in each of six paper or foil jelly cases. Beat egg whites to soft peaks and gradually beat in sugar until stiff peaks form. In another bowl, beat yolks until thick and beat in lemon peel and juice until well mixed. Whip cream lightly, then fold egg yolk mixture and cream into egg whites until just mixed. Put mixture into cases and sprinkle with more cornflake crumbs.
Pack by putting foil lid on each dish.
To serve thaw in refrigerator for 30 minutes.
Storage Time 1 month.

FRUIT MOUSSE

¼ pint fruit purée	2 egg whites
1 oz. caster sugar	Juice of ½ lemon
¼ pint double cream	

Mix fruit purée and sugar. Whip cream lightly, and whip egg whites stiffly. Add lemon juice to fruit, then fold in cream and egg whites. A little colouring may be added if the fruit is pale.
Pack in serving dish covered with foil.
To serve thaw in refrigerator without lid for 2 hours.
Storage Time 1 month.

FRUIT CREAM

1 lb. raspberries, currants, gooseberries or blackberries	6 oz. sugar
¾ pint water	2 tablespoons cornflour

Clean fruit. Bring water to boil, add fruit and sugar, and boil until fruit is soft. Mix cornflour with a little cold water, blend into hot liquid, and bring back to boil. Cool.
Pack in serving dish covered with foil.
To serve thaw in refrigerator for 1 hour and serve with cream.
Storage Time 1 month.

PEARS IN RED WINE

8 eating pears
8 oz. sugar
¼ pint water
¼ pint burgundy
2 ins. cinnamon stick

Peel pears but leave whole with stalks on. Dissolve sugar in water and add cinnamon stick. Simmer pears in syrup with lid on for 15 minutes, then add burgundy and uncover the pan. Continue simmering for 15 minutes. Drain pears and put into individual leak-proof containers. Reduce syrup by boiling until it is thick, then pour over pears, and cool.

Pack in leak-proof containers since the syrup does not freeze solid; the pears lose moisture on thawing and thin the syrup, but the effect is lessened if they are packed in individual containers.

To serve thaw in refrigerator for 8 hours.
Storage Time 2 months.

PEACHES IN WHITE WINE

8 peaches
½ pint white wine
8 oz. sugar
1 tablespoon Kirsch

Peel peaches and cut in halves. Put into oven dish, cut sides down, cover with wine and sprinkle with sugar. Bake at 375°F (Gas Mark 5) for 40 minutes. Stir in Kirsch and cool.

Pack in leak-proof containers, allowing two peach halves to each container, and covering with syrup.

To serve heat at 350°F (Gas mark 4) for 45 minutes, adding a little more Kirsch if liked, and serve with cream.
Storage Time 2 months.

ICEBOX CAKE

6 oz. icing sugar
4 oz. butter
2 medium eggs

2 teaspoons grated lemon peel and 2 tablespoons lemon juice
or 2 tablespoons cocoa and 1 teaspoon coffee essence.
48 sponge finger biscuits.

Cream sugar and butter until light and fluffy and work in eggs one at a time. Gradually beat in flavourings, and then beat hard until fluffy and smooth. Cover a piece of cardboard with foil and on it place 12 biscuits, curved side down. On this put one-third of the creamed mixture. Put another layer of biscuits in opposite direction, and more creamed mixture. Repeat layers, ending with biscuits.

Pack by wrapping in foil. This is a large pudding and could be prepared in two portions.

To serve unwrap and thaw in refrigerator for 3 hours, then cover completely with whipped cream and serve at once.

Storage Time 1 month.

COFFEE PUDDING

3 oz. caster sugar
4 oz. butter
4 oz. fresh white fine breadcrumbs
5 tablespoons strong black coffee

Cream butter and sugar until light and fluffy. Work in breadcrumbs and coffee until completely mixed. Press into dish.

Pack by covering serving dish with foil lid.

To serve thaw uncovered in refrigerator for 45 minutes, cover with whipped cream and decorate with nuts.

Storage Time 1 month.

BAKED CHEESE CAKE

2 oz. digestive biscuit crumbs
1 lb. cottage cheese
1 teaspoon lemon juice
1 teaspoon grated orange rind
1 tablespoon cornflour
2 tablespoons double cream
2 eggs
4 oz. caster sugar

Use an 8 ins. cake tin with removable base to bake this cheesecake. Butter sides and line base with buttered paper. Sprinkle with crumbs. Sieve cottage cheese and mix with lemon juice, orange rind and cornflour. Whip cream and stir in. Separate eggs, and beat egg yolks until thick, then stir into cheese mixture. Beat egg whites until stiff and beat in half the sugar, then stir in remaining sugar. Fold into cheese mixture and put into baking tin. Bake at 350°F (Gas Mark 4) for 1 hour, and leave to cool in the oven. Remove from tin.

Pack in foil *after* freezing, and then in box to avoid crushing.

To serve thaw in refrigerator for 8 hours.

Storage Time 1 month.

SWEDISH APPLECAKE

3 oz. fresh brown breadcrumbs 2 tablespoons brown sugar
1 oz. butter 1 lb. apples

Gently fry breadcrumbs in butter until golden brown. Cook apples in very little water until soft, and sweeten to taste. Stir brown sugar into buttered crumbs. Arrange alternate layers of buttered crumbs and apples in buttered dish, beginning and ending with a layer of crumbs. Press firmly into dish and cool.
Pack by covering with foil lid.
To serve thaw without lid in refrigerator for 1 hour, turn out and serve with cream.
Storage Time 1 month.

BAKED APPLE DUMPLINGS

8 apples Butter
Sugar
8 oz. short pastry

Core apples, leaving $\frac{1}{4}$ ins. core at bottom of each to hold filling. Fill with sugar and put a knob of butter in each. Put each apple on a square of pastry and seal joins. Bake at 425°F (Gas Mark 7) for 25 minutes. Cool.
Pack in containers or foil dishes and cover with foil lids.
To serve put in oven while still frozen and heat at 375°F (Gas Mark 5) for 20 minutes, serving with cream, custard or hot apricot jam.
Storage Time 1 month.

Chapter Six

PIES, PASTRY AND PIE FILLINGS

Pies which are frozen provide useful emergency meals; they are also a neat way of storing surplus fruit and using leftover meat and poultry. Not only large pies can be stored, but a variety of turnovers, pasties and individual fruit pies.

Short pastry and flaky pastry freeze equally well either cooked or uncooked, but a standard balanced recipe should be used for best results. Unbaked pastry will store up to 4 months and baked pastry up to 6 months, but this depends on the filling of the pies.

Pies and flans may be stored baked and unbaked. The baked pie stores for a longer period, but the unbaked pie has a better flavour and scent, and the pastry is crisper and flakier. Almost all fillings can be used, except those with custard which separates. Meringue toppings toughen and dry during storage.

UNBAKED PASTRY

Pastry may be rolled, formed into a square, wrapped in greaseproof paper, and then in foil or polythene for freezing. This pastry takes time to thaw, and may crumble when rolled. It should be thawed slowly, then cooked as fresh pastry and eaten fresh-baked, not returned to the freezer in cooked form.

BAKED PASTRY

Flan cases, patty cases and vol-au-vent cases are all useful if ready baked. For storage, it is best to keep them in the cases in which they are baked or in foil cases. Small cases may be packed in boxes in layers with paper between. Baked cases should be thawed in wrappings at room temperature before filling. A hot filling may be used and the cases heated in a low oven.

UNBAKED PIES

Pies may be prepared with or without a bottom crust. Air vents in pastry should not be cut before freezing. To prevent sogginess, it is better to freeze unbaked pies before wrapping them. To bake pies, cut slits in top crust and bake unthawed as for fresh pies, allowed about 10 minutes longer than normal cooking time.

BAKED PIES

Pies may be baked in the normal way, then cooled quickly before freezing. A pie is best prepared and frozen in foil, but can be stored in a rustproof and crack-proof container. The container should be put into freezer paper or polythene for freezing. A cooked pie should be heated at 375°F (Gas Mark 5) for 40–50 minutes for a double-crust pie, 30–50 minutes for a one-crust pie, depending on size. Cooked pies may also be thawed in wrappings at room temperature and eaten without reheating.

FRUIT FILLINGS

If the surface of the bottom crust of fruit pies is brushed with egg white, it will prevent sogginess. Fruit pies may be made with cooked or uncooked fillings. Apples tend to brown if stored in a pie for more than 4 weeks, even if treated with lemon juice, and it is better to combine frozen pastry and frozen apples to make a pie.

If time is short, it is convenient to freeze ready-made fruit pie fillings, ready to fit into fresh pastry when needed, and this is a good way of freezing surplus fruit in a handy form. The mixture is best frozen in a spongecake tin or an ovenglass pie plate lined with foil, then removed from container and wrapped in foil for storage; the same container can then be used for making a pie at a later date. A little cornflour or flaked tapioca gives a firm filling which cuts well and does not seep through the pastry.

MEAT FILLINGS

Meat pies may be completely cooked so that they need only be reheated for saving. Preparation time is saved however if the meat filling is cooked and cooled, then topped with pastry. If the pie is made in this form, the time taken to cook the pastry is enough to heat the meat filling, and this is little longer than heating the whole pie.

Pies are most easily frozen in foil containers which can be used in the oven for final cooking. If a bottom crust is used, sogginess will be prevented if the bottom pastry is brushed with melted butter or lard just before filling. Pies should be reheated at 400°F for required time according to size, and are best stored no longer than 2 months.

HOT WATER CRUST PIES

These are normally eaten cold and can be frozen baked or unbaked, but there are obvious risks attached to freezing them. The pastry is made with hot water, and the pie may be completely baked, and cooled before freezing; however the jelly must be added just before the pie is to be served. The easiest way to do this is to freeze the stock separately at the time of making the pie, and when the pie is thawing (which takes about 4 hours), the partially thawed pie can be filled with boiling stock through the hole in the crust, and this will speed up the thawing process.

The second method involves freezing the pie unbaked, partially thawing and then baking. However, this means the uncooked meat will be in contact with the warm uncooked pastry during the making process, and unless the pie is very carefully handled while cooling, there is every risk of dangerous organisms entering the meat. On balance, it would seem preferable to avoid freezing this type of pie.

OPEN TARTS

Tarts with only a bottom crust may be filled and frozen. They are better frozen before wrapping to avoid spoiling the surface of the filling during packing.

STEAK AND KIDNEY PIE

1 lb. steak	Salt and pepper
4 oz. kidney	1 tablespoon cornflour
¾ pint stock	8 oz. short or flaky pastry

Cut steak and kidney into neat pieces and fry until brown in a little dripping. Add stock and seasoning and simmer for 2 hours. Thicken gravy with cornflour and pour mixture into foil dish. When meat is cold, cover with pastry, pack and freeze.

Pack in foil container with foil lid, or put container inside polythene bag.

To serve bake at 400°F (Gas Mark 6) until pastry is cooked and golden.

Storage Time 1 month.

CHICKEN PIE

5 lb. boiling chicken	1 lb. carrots
2 celery stalks	2 lbs. shelled peas
1 medium onion	6 oz. mushrooms
½ sliced lemon	½ pint thin cream
2 sprigs parsley	Pinch of nutmeg
1 bayleaf	2 oz. cornflour
Salt and pepper	2 lbs. flaky pastry

Simmer chicken in water for 2½ hours with celery, onion, lemon, parsley, bayleaf, salt and pepper. Cool chicken in stock and cut flesh from bones in neat cubes. Slice carrots and cook carrots and peas for 15 minutes. Cook sliced mushrooms in a little butter. Drain vegetables and mix with chicken flesh. Measure out 2 pints of chicken stock and make a sauce with cornflour, a seasoning of nutmeg, salt and pepper to taste, and stir in the thin cream without boiling. Simmer for 3 minutes until smooth, pour over chicken mixture and cool completely.

Divide mixture into foil pie plates and cover with flaky pastry. This quantity of filling will make eight 6 ins. diameter pies.
Pack by wrapping containers in foil or polythene bags.
To serve cut slits in pastry and put dishes on baking sheet. Bake at 450°F (Gas Mark 8) for 40 minutes.
Storage Time 2 months.

PIGEON PIE

6 pigeons	Salt, pepper and mace
8 oz. steak	8 oz. short pastry

Optional – Small mushrooms and/or hard-cooked egg yolks
Remove breasts from pigeons with a sharp knife and put into a saucepan with the steak cut into small pieces. Season with salt, pepper and a pinch of mace and just cover with water. Simmer with lid on for 1 hour. If liked, mix with mushrooms tossed in a little butter, or egg yolks. Cool completely. Put into foil baking dish and cover with pastry.
Pack by covering container with foil, or putting into polythene bag.
To serve cut slits in pastry and bake at 400°F (Gas Mark 6) for 45 minutes.
Storage Time 2 months.

CORNISH PASTIES

1 lb. short pastry	1 onion
12 oz. steak	3 tablespoons stock
6 oz. potatoes	Salt and pepper

Cut steak and potatoes in small dice and chop onion finely. Season and moisten with stock. Divide pastry into eight pieces and roll these into 5 ins. circles. Put meat fillings in the centre of each circle of pastry, and fold up edges to make half-circles. Seal edges well, giving a fluted appearance, and slightly flatten base of pasties.

Pack in polythene bags; if pasties are baked before freezing, pack in boxes to avoid crushing.

To serve unbaked pasties, brush them with egg and bake (without thawing) at 425°F (Gas Mark 7) for 15 minutes, then at 350°F (Gas Mark 4) for 40 minutes. If preferred, pasties may be baked at the same temperature and for the same time before freezing, then thawed for 12 hours in a refrigerator to eat cold or reheated at 375°F (Gas Mark 5) for 20 minutes.

Storage Time 2 months.

FISH TURNOVERS

8 oz. flaky pastry	4 tomatoes
8 oz. cooked haddock or cod	1 teaspoon curry powder
1 oz. butter	Salt and pepper

Roll pastry into two 12 ins. squares. Flake the fish and mix with melted butter, curry powder, salt and pepper. Divide mixture between two pieces of pastry. Skin tomatoes, and cover fish mixture with tomato slices. Fold in corners of pastry to form envelope shapes and seal edges.

Pack in foil or polythene bags.

To serve put frozen turnovers on tray and bake at 475°F (Gas Mark 9) for 20 minutes, then at 400°F (Gas Mark 6) for 20 minutes.

Storage Time 1 month.

RASPBERRY AND APPLE PIE FILLING

8 ozs. thinly sliced cooking apples	8 oz. sugar
1 lb. raspberries	2 tablespoons tapioca flakes
1 tablespoon lemon juice	Pinch of salt

Mix all ingredients well in a bowl and leave to stand for 15 minutes. Line a pie plate with foil, leaving 6 ins. rim. Put filling into foil, fold over and freeze. Remove frozen filling from pie plate, pack and freeze. Other good combinations are rhubarb and orange, or apricot and pineapple. Single fruits such as cherries and blackberries may be prepared in the same way.

Pack in foil or polythene.

To serve line pie plate with pastry, put in frozen filling, dot with butter, cover with pastry lid, make slits in top crust, and bake at 425°F (Gas Mark 7) for 45 minutes.

Storage Time 12 months.

BLACKCURRANT FLAN

8 oz. plain flour	1 small egg
1 teaspoon cinnamon	1½ teaspoons lemon juice
5 oz. butter	1½ lbs. fresh blackcurrants
1½ oz. ground almonds	6 oz. sugar
1½ oz. caster sugar	

Mix flour and cinnamon and work in butter until mixture is like fine crumbs. Mix in almonds and caster sugar and make into a paste with the egg and lemon juice. Divide mixture to make two flans and line flan rings or foil cases, reserving some pastry for decoration (this pastry is very delicate to handle). Put half the prepared fruit in each flan case and sprinkle evenly with sugar. Cover flans with pastry lattice. The flans may be frozen uncooked or cooked, and are less likely to be soggy if frozen before baking.

Pack in foil, or put foil cases into polythene bags.

To serve brush lattice of unbaked flan with water, sprinkle with caster sugar and put in cold oven set at 400°F (Gas Mark 6), and bake for 45 minutes. The flan may be baked before freezing at 400°F (Gas Mark 6) for 30 minutes, then thawed in loose wrappings at room temperature for 3 hours.

Storage Time 2 months.

BAKEWELL TART

8 oz. puff pastry	4 oz. sugar
4 oz. jam	2 eggs
4 oz. butter	4 oz. ground almonds

Line pie plate or foil dish with pastry and spread with jam. Melt the butter, stir in sugar, then beaten eggs and almonds. Beat well and put over jam. Bake at 350°F (Gas Mark 4) for 40 minutes. Cool.
Pack in foil or polythene bags
To serve thaw in loose wrappings at room temperature for 3 hours.
Storage Time 2 months.

CUMBERLAND TART

8 oz. short pastry	4 oz. plain flour
½ pint thick sweet applesauce	*Icing*
2 oz. butter	2 oz. butter
1 oz. sugar	2 oz. icing sugar
1 egg	

Line baking tin with pastry and cover with applesauce. Cream butter and sugar, beat in egg, and add flour. Put on top of applesauce and bake at 400°F (Gas Mark 6) for 30 minutes. Leave until cold. Cream butter and icing sugar until light and fluffy and spread on top of tart. Freeze before wrapping.
Pack in foil or polythene bag.
To serve remove wrappings and thaw at room temperature for 3 hours.
Storage Time 2 months.

DANISH CHERRY TART

8 oz. short pastry	6 oz. icing sugar
8 oz. stoned cooking cherries	2 eggs
4 oz. ground almonds	

Line a pie plate or foil dish with pastry and prick the pastry well. Fill with cherries. Mix ground almonds, sugar and eggs one at a time to a soft paste. Pour over cherries and bake at 400°F (Gas Mark 6) for 25 minutes. Cool.
Pack in foil or polythene bags
To serve thaw in wrappings at room temperature for 3 hours.
Storage Time 2 months.

Chapter Seven

PASTA DISHES

Pasta such as spaghetti and macaroni, together with rice, may be successfully frozen to be used later with a variety of sauces which may also be frozen. Additionally, composite meals featuring pasta may be prepared and frozen, and are extremely useful luncheon or supper dishes. A quantity of pasta shapes, or rice, may also be frozen to use with soup; if frozen in liquid, these starchy items become slushy, but can conveniently be added to soup during the reheating period.

TO FREEZE SPAGHETTI, MACARONI OR RICE
The pasta should be slightly undercooked in boiling salted water, after thorough draining, it should be cooled under cold running water in a sieve, then shaken as dry as possible, packed into polythene bags, and frozen. To serve, the pasta is put into a pan of boiling water and brought back to the boil, then simmered until just tender, the time depending on the state in which it has been frozen. Rice may also be reheated in a frying pan with a little melted butter. Composite dishes can be reheated in a double boiler, or in the oven under a foil lid.

SAUCES
Sauces made with flour are likely to curdle when reheated, and cornflour should be used for thickening if necessary. Large quantities of sauce may be stored in waxed or rigid plastic containers, or frozen in "brick" form in loaf tins, then stored in foil in the freezer. Since small quantities of sauce are often needed for individual portions, it can also be frozen in ice cube trays, then each cube wrapped in foil for storage.

CHEESE
Many pasta dishes require the addition of grated cheese. This may be prepared and frozen in small plastic bags, and is a useful way of using up odd ends of cheese.

TOMATO SAUCE

1 lb. tomatoes	1 oz. ham
1 oz. butter	1 pint stock
1 small onion	Parsley, thyme, bay leaf
1 small carrot	1 oz. cornflour.

Cut tomatoes in slices. Melt butter and fry sliced onion and carrot until golden. Add tomatoes, ham, stock and herbs and simmer for 30 minutes. Put through a sieve, thicken with cornflour, season again to taste, and cook for 5 minutes stirring well. Cool.

Pack in containers, in brick form, or in ice-cube trays, wrapping bricks or cubes when frozen in foil.

To serve thaw in double boiler, stirring gently.

Storage Time 12 months.

SPAGHETTI SAUCE

1 large chopped onion	6 oz. tomato purée
1 clove garlic	½ pint water
2 tablespoons olive oil	1 teaspoon salt
1 lb. minced beef	½ teaspoon pepper
1 lb. chopped peeled tomatoes	1 bay leaf

Fry onion and crushed garlic in oil, add beef and cook until browned. Add all the other ingredients and simmer slowly for 1 hour until thick and well blended. Cool.

Pack in containers, in brick form, or in ice cube trays, wrapping frozen bricks or cubes in foil for storage.

To serve thaw gently over direct heat.

Storage Time 2 months.

CURRY SAUCE

2 medium onions	1 tablespoon vinegar
1 tablespoon curry powder	1 tablespoon chutney
1 tablespoon cornflour	1 tablespoon sultanas
1 pint stock	

Slice onions and fry in a little butter until soft. Add curry powder and cook for 1 minute. Gradually add stock and cornflour blended with a little water. Add remaining ingredients and simmer 15 minutes. Cool. This sauce may also be frozen with the addition of some small chicken, lamb or beef pieces, but should then only be stored for 2 months.

Pack in containers, in brick form or ice cube trays, wrapping frozen bricks or cubes in foil for storage.

To serve reheat in a double boiler, stirring gently.

Storage Time 6 months.

MACARONI CHEESE

8 oz. macaroni	10 oz. cheese
4 oz. butter	1 teaspoon salt
4 tablespoons flour	Pepper
1½ pints milk	

Optional – chopped ham, chopped cooked onions or mushrooms.

Cook macaroni as directed and drain well. Melt butter, blend in flour and work in milk, cooking to a smooth sauce. Over low heat, stir in cheese and seasoning. Mix macaroni and cheese and cool thoroughly.

Pack into foil-lined casserole, and remove solid block when frozen bricks or cubes in foil for storage.

To serve turn into casserole, cover with foil and heat at 400°F (Gas Mark 6) for 1 hour, removing foil for the last 15 minutes to brown top.

Storage Time 2 months.

CHICKEN TETRAZZINI

2 lbs. cooked chicken meat	½ pint creamy milk
12 oz. spaghetti	Salt and pepper
1 lb. small mushrooms	Pinch of mace
4 oz. butter	4 tablespoons sherry
2 oz. plain flour	3 oz. grated cheese
1 pint chicken stock	

Cut the chicken meat into small thin strips. Slightly undercook the spaghetti in boiling water, and drain thoroughly. Slice mushrooms thinly and cook gently in half the butter. Melt remaining butter, work in flour, and gradually add stock, then bring back to the boil, stir in creamy milk and cook very gently for 10 minutes. Add salt, pepper and mace, and stir in sherry and cheese. Moisten the chicken meat with a little sauce, and cool. Use rest of sauce to mix together spaghetti and mushrooms. arrange spaghetti mixture in a foil-lined casserole and top with chicken mixture.

Pack by wrapping foil block in more foil for storage.

To serve return to casserole and bake at 375°F (Gas Mark 5) for 1½ hours, sprinkling with a little grated cheese thirty minutes before serving time to give a brown top.

Storage Time 2 months.

SIMPLE RISOTTO

1 onion	2 pints chicken stock
2½ oz. butter	1 oz. cheese
12 oz. rice	Salt and pepper

Chop onion finely and cook in half the butter until soft and transparent. Add the rice and cook, stirring well, until it is buttered but not brown. Add stock which has been heated very gradually, adding more until it has been absorbed. Cook over a low heat for about 30 minutes until the liquid has been taken up and the rice is soft but firm. Stir in cheese, remaining butter, salt and pepper, and cool. This mixture may be frozen on its own, or with the addition of cooked peas, mushrooms, flaked fish or shellfish, chopped ham or chicken, or chicken livers, which should be added during the last 10 minutes' cooking.

Pack in waxed or rigid plastic container, or polythene bags.

To serve reheat gently over direct heat, and serve sprinkled with grated cheese.

Storage Time 2 months.

SPANISH RICE

8 oz. Patna rice	6 oz. mushrooms
2 oz. oil	8 oz. tomatoes
4 oz. lean bacon	Salt and pepper
12 oz. onions	

Slightly undercook rice in boiling salted water and drain well. Heat oil and cook chopped bacon and onions gently for 10 minutes. Add sliced mushrooms and peeled and sliced tomatoes and continue cooking until tender and well blended. Stir in rice and seasoning and cool.

Pack in waxed or rigid plastic containers, or in polythene bags.

To serve reheat gently over direct heat, or use to fill green peppers, aubergines or marrow.

Storage Time 2 months.

GNOCCHI

1 onion	1½ oz. grated cheese
1 bay leaf	½ oz. butter
1 pint milk	1 teaspoon French mustard
5 tablespoons polenta or semolina	*Topping*
	2 oz. butter
Salt and pepper	2 oz. grated cheese

Put onion and bay leaf in milk and bring slowly to the boil with the lid on the pan. Remove onion and bay leaf, stir in polenta or semolina and mix well. Season with salt and pepper and simmer 15 minutes until creamy. Remove from fire, stir in cheese, butter and mustard, and spread out on a tin about ¾ ins. thick. When cold and set, cut in squares.

Pack in overlapping layers in foil tray, brush with melted butter and grated cheese, and cover with foil, or put in polythene bag.

To serve thaw at room temperature for 1 hour, then bake at 350°F (Gas Mark 4) for 45 minutes until golden and crisp.

Storage Time 2 months.

Chapter Eight

ICES AND FROZEN PUDDINGS

Home-made ice cream can be very successfully made and stored in the freezer. Special crank attachments can now be bought for the home freezer which beat the ice cream as it freezes and produce a really smooth product; these are however expensive and will take time to repay themselves, and most cooks will prefer the slightly rougher home-made version.

Bought ice cream is less expensive in bulk and will store well, and may be made into a number of puddings by those who do not wish to make their own ices. Large quantities may be divided into meal-sized portions for storage, or be converted into ready-to-serve frozen puddings. If the ice cream is left in a large container after portions have been removed, the surface should be covered with foil to retain flavour and texture and to avoid ice crystals forming.

Home-made ice cream is best made with pure cream and gelatine or egg yolks. Evaporated milk may be used, but the flavour is not so good (the unopened tin of milk should be boiled for 10 minutes, cooled and left in a refrigerator overnight before using). Eggs, gelatine, cream or sugar syrup are all good emulsifying agents and will help to give smoothness and prevent large ice crystals forming; gelatine gives a particularly smooth ice. Whipped egg whites give lightness. Too much sugar prevents freezing, but sweetness will diminish in the freezer, so a well-balanced recipe must be used.

BASIC METHOD

Ice cream must be frozen quickly, or it will be "grainy" and will retain this rough texture during storage. Whatever emulsifying agent is used, the method is the same. The mixture should be packed into trays and frozen until just solid about ½ ins. from the edge; then beaten quickly in a chilled bowl and frozen for a further hour. This freezing followed by beating should be repeated for up to three hours. It is possible to pack the basic mixture into storage containers and freeze after the first beating, but while this saves time, the result is not so smooth, and it is preferable to complete the ice cream before packing for storage.

BASIC FLAVOURINGS

Flavourings should be strong and pure (e.g. vanilla pod or sugar instead of essence; liqueurs rather than flavoured essences), as they are affected by low temperature storage. Flavourings may be varied by using one of the basic recipes and adjusting to the required flavour.

Butterscotch – Cook the sugar in the recipe with 2 tablespoons butter until well browned, then add to hot milk or cream.

Caramel – Melt half the sugar in the recipe with a moderate heat, using a heavy saucepan, and add slowly to the hot milk.

Chocolate – Melt 2 oz. unsweetened cooking chocolate in 4 tablespoons hot water, stir until smooth, and add to the hot milk.

Coffee – Scald 2 tablespoons ground coffee with milk or cream and strain before adding to other ingredients.

Peppermint – Use oil of peppermint, and colour lightly green.

Praline – Make as caramel flavouring, adding 4 oz. blanched, toasted and finely chopped almonds.

Egg Nog – Stir in several tablespoons rum, brandy or whisky to ice cream made with egg yolks.

Ginger – Add 2 tablespoons chopped preserved ginger and 3 tablespoons ginger syrup to basic mixture.

Maple – Use maple syrup in place of sugar and add 4 oz. chopped walnuts.

Pistachio – add 1 teaspoon almond essence and 2 oz. chopped pistachio nuts, and colour lightly green.

MIXED FLAVOURINGS

Mixed flavour ice creams can be prepared by adding flavoured sauces or crushed fruit to vanilla ice cream, or by making additions to some of the basic flavours. Crushed fruit such as strawberries, raspberries or canned mandarin oranges may be beaten into vanilla ice cream before packing. Chocolate or butterscotch sauce can be swirled through vanilla ice. Chopped toasted nuts or crushed nut toffee pair with vanilla, coffee or chocolate flavours. A pinch of coffee powder may be used in chocolate ice cream, or a little melted chocolate in coffee ice; one of the chocolate- or coffee-flavoured liqueurs may also be used.

ICE CREAM BOMBES

Ice cream bombes or moulds are both decorative and delicious, and are a good way of using small quantities of rather special flavours. They may be made with both bought or home-made ice cream, or with a combination of both. Any metal mould or bowl may be used, or special double-sided moulds may be purchased; metal jelly moulds are excellent for the purpose, and may be covered with foil for storage.

The ice cream should be slightly softened before using in a mould, and pressed firmly into the container with a metal spoon; each layer should be pressed down and frozen before a second layer is added or the flavours and colours may run into each other. Moulds may also be lined with one flavour ice cream, and the centre filled with another flavour, or with mixtures of fruit, liqueurs and ice cream.

Bombes are unmoulded by turning on to a chilled plate and covering the mould with a cloth wrung out in hot water, shaking slightly to release the ice cream. It is a good idea to unmould the bombe about an hour before serving, then wrap the ice cream in foil and return to the freezer to keep firm.

BOMBE FLAVOURINGS

Melon Bombe – Line mould with pistachio ice cream and freeze 30 minutes. Fill with raspberry ice cream mixed with chocolate chips and wrap for storage. This looks like a watermelon when cut in slices.

Raspberry Bombe – Line mould with raspberry ice cream, freeze for 30 minutes and fill with vanilla ice cream.

Coffee Bombe – Line mould with coffee ice cream, freeze 30 minutes. Fill with vanilla ice cream flavoured with chopped maraschino cherries and syrup.

Three-Flavour Bombe – Line mould with vanilla ice cream, freeze 30 minutes. Put in thin lining of praline ice cream and freeze 30 minutes. Fill centre with chocolate ice cream.

Tutti Frutti Bombe – Line mould with strawberry ice cream, freeze 30 minutes. Fill with lemon sorbet mixed with drained canned fruit cocktail.

Raspberry-Filled Bombe – Line mould with vanilla or praline ice cream and freeze 30 minutes. Fill with crushed fresh raspberries beaten into whipped cream and lightly sweetened.

Peach-Filled Bombe – Line mould with vanilla ice cream and freeze 30 minutes. Fill with whipped cream mixed with chopped drained canned peaches, lightly sweetened and flavoured with light rum.

CUSTARD ICE

¾ pint creamy milk
1 vanilla pod
2 large egg yolks
2 oz. sugar
Small pinch of salt
½ pint thick cream

Scald milk with vanilla pod. Remove pod and pour milk on to egg yolks lightly beaten with sugar and salt. Cook mixture in a double boiler until it coats the back of a spoon. Cool and strain and stir in the cream. Pour into freezing trays and beat twice during a total freezing time of about 3 hours. Pack into containers, cover and seal, and store in freezer.

GELATINE ICE

¾ pint creamy milk
1 vanilla pod
1 dessertspoon gelatine
3 oz. sugar
Pinch of salt

Heat ¼ pint milk with vanilla pod to boiling point. Soak gelatine in 2 tablespoons cold water, then put into a bowl standing in hot water until the gelatine is syrupy. Pour warm milk on to the gelatine, stir in sugar, salt and remaining milk. Remove vanilla pod and freeze mixture, beating twice during 3 hours total freezing time. Pack into containers, cover and seal, and store in freezer. This mixture is particularly good for using with flavourings such as chocolate or caramel.

CREAM ICE

1 pint thin cream
1 vanilla pod
3 oz. sugar
Pinch of salt

Scald cream with vanilla pod, stir in sugar and salt, and cool. Remove vanilla pod and freeze mixture to a mush. Beat well in a chilled bowl, and continue freezing for 2 hours, beating once more. Pack into containers, cover and seal, and store in freezer.

ORANGE SORBET

2 teaspoons gelatine	1 teaspoon grated orange rind
½ pint water	½ pint orange juice
6 oz. sugar	4 tablespoons lemon juice
1 teaspoon grated lemon rind	2 egg whites

Soak gelatine in a little of the water and boil the rest of the water and sugar for 10 minutes to a syrup. Stir gelatine into syrup and cool. Add rinds and juices. Beat egg whites until stiff but not dry and fold into mixture. Freeze to a mush, beat once, then continue freezing, allowing 3 hours total freezing time (This ice will not go completely hard). Pack into containers, cover and seal, and store in freezer.

Lemon Sorbet. Follow the same method, using only lemon juice and rind to make up the total quantities, instead of a mixture of orange and lemon.

Fresh Fruit Cases. Orange and lemon sorbets are particularly attractive if packed into fruit skins. Scoop out the oranges or lemons, wash thoroughly and dry and pack in sorbet when it is ready for storage, leaving surface raised above fruit skins. Wrap containers in foil for storage. If this is not done before storage, the skins may be prepared and left wet and the sorbet packed in and returned unwrapped to the freezer for 1 hour before serving.

FRESH FRUIT ICE

¾ pint cream	1½ tablespoons caster sugar
½ pint fruit purée	

Beat cream lightly until thick, stir in fruit purée and sugar, and pour into freezer tray. Freeze without stirring. Put into containers, cover and seal for storage. This is very good with fresh raspberries, or with apricots poached in a little vanilla-flavoured syrup before sieving.

APPLE ICE CREAM

½ pint apple sauce
Pinch of nutmeg and
   cinnamon
2 teaspoons lemon juice
Sugar
½ pint cream

Sieve apple sauce, add spices and lemon juice and sweeten very lightly to taste. Chill for 1 hour. Fold in whipped cream and pour into freezing tray. Freeze until firm. Pack in containers, cover and label for storage. This ice is very popular with children and is delicious wth butterscotch or raspberry sauce.

STRAWBERRY WATER ICE

8 oz. strawberries
Juice of 1 orange
¾ pint water
12 oz. sugar
1 egg white

Crush strawberries with orange juice and put through sieve. Put water in pan, stir in sugar and boil for 5 minutes. Cool, and stir in strawberries. Freeze to a mush, then put into chilled bowl. Beat well and add stiffly-whipped egg white. Freeze until firm. Pack into containers, cover and label for storage.

RUM AND DATE ICE CREAM

3 oz. chopped dates
2 oz. chopped walnuts
1 tablespoon dark rum
1 pint vanilla ice cream

Mix dates, walnuts and rum and leave to stand for 30 minutes. Slightly soften ice cream and blend in rum mixture. Freeze until firm, pack into containers or mould, cover and seal for storage.

CHOCOLATE PUDDING ICE

2 oz. halved stoned raisins
1½ oz. currants
½ oz. candied orange peel
1 oz. halved glacé cherries
Liqueur glass rum or brandy
1 pint chocolate ice cream

Soak raisins, currants, peel and cherries in rum or brandy overnight. Fold into slightly softened ice cream and pack into metal pudding mould. Cover and seal for storage. This may be used as a substitute for the traditional Christmas pudding for parties, and is good served with liqueur-flavoured cream.

ICE CREAM LAYER CAKE

2 pints strawberry ice cream
2 pints vanilla ice cream
8 tablespoons strawberry jam
1 lb. strawberries
1 pint double cream

Slightly soften ice creams and press into sponge cake tins to make two strawberry layers and 2 vanilla layers. Cover each tin with foil and freeze until firm. Unmould ice cream and arrange in alternate layers spread with strawberry jam (this is best done on a foil-covered cake board). Cover with whipped cream and strawberries, and freeze uncovered for 3 hours. Wrap in foil for storage. The cake may be made without the cream and strawberries which can be added just before serving. Fresh strawberries may also be chopped and added to the strawberry ice cream before moulding in the spongecake tins.

LEMON ICE PIE

8 oz. digestive biscuit crumbs
2 oz. sugar
2 oz. melted butter
½ pint vanilla ice cream
1 pint lemon sorbet

Mix crumbs, sugar and melted butter and press into flan ring or foil dish. Freeze until firm, keeping back about 2 tablespoons of the buttered crumbs for garnishing. In the pie shell, place scoops of vanilla ice cream, then top with scoops of lemon ice and scatter with reserved crumbs. Return to freezer until firm, wrap in foil, seal and store.

Chapter Nine

CAKES AND BISCUITS, ICINGS AND FILLINGS

Cakes and biscuits may be frozen in both cooked and uncooked forms; they freeze extremely well and cakes, particularly, taste fresher than when stored in tins. Filled and iced cakes may also be successfully frozen, saving time for parties and special occasions, since these cakes cannot be stored in tins.

INGREDIENTS
Good fresh ingredients must be used for cakes which are to be frozen. Stale flour deteriorates quickly after freezing. Butter cakes have the best flavour, but margarine may be used for strongly flavoured cakes such as chocolate, and gives a good light texture. Eggs must be fresh and well-beaten, as yolks and whites freeze at different speeds and will affect the texture of the cake.

Boiled icings and those made with cream or egg whites will crumble on thawing, and the best icings for freezing are those made with butter and icing sugar. Fruit fillings and jams will make a cake soggy.

FLAVOURINGS AND DECORATIONS
Synethetic flavourings develop off-flavours during freezing; this is particularly the case with vanilla. They must therefore always be pure (e.g. vanilla pod or vanilla sugar should be used). Highly spiced foods also develop off-flavours and spice cakes are best not frozen, although basic gingerbread is satisfactory for a short storage time. Chocolate, coffee and fruit-flavoured cakes freeze very well.

There is little advantage in decorating cakes before they are frozen as there will be plenty of time to do this while they are thawing; otherwise moisture may be absorbed by the decorations and colour changes will affect the appearance of the cake.

Decorations such as nuts, grated chocolate and coloured balls can be added when the cake is fully thawed.

ICINGS AND FILLINGS

Cakes for freezing should not be filled with cream, jam or fruit. Butter icings are best, but an iced cake must be absolutely firm before wrapping and freezing. Cakes may also be frozen before wrapping, and this can be very convenient to avoid spoiling the surface of soft icing.

Wrappings must be removed before thawing to allow moisture to escape, and to avoid smudging icing.

If sponge or flavoured cakes are to be iced after thawing, they can be packed in layers with Cellophane, foil or greaseproof paper between them, and the layers can be separated easily for filling and icing when thawed.

PACKAGING

Small cakes can be frozen in polythene bags in convenient quantities; small iced cakes are better packed in boxes to avoid crushing or smudging. Large quantities of small iced cakes can be frozen on baking trays unwrapped, then packed in layers in boxes with Cellophane or greaseproof paper between.

Large cakes can be frozen in polythene bags or in heavy duty foil. It is also possible to pack individual pieces of cake for lunch-boxes; these pieces may be frozen individually in bags or boxes, but it is easier to slice the whole cake in wedges before freezing, and to withdraw slices as they are needed without thawing the whole cake.

SPONGECAKES

Fatless sponges can be stored for 10 months, but those made with fat store for 4 months. Unbaked sponges will keep for 2 months, but lose volume in cooking; they should be prepared in rustless baking tins or foil. Cake batter can be stored in cartons if this is easier for packing; it should be thawed in the container before putting into baking tins, but if the mixture thaws too long, the cake will be heavy. Delicate baked cakes may be frozen in paper or polythene, then packed in boxes to avoid crushing. All baked cakes should be thawed in wrappings at room temperature unless they are iced, when the wrappings should be removed before thawing.

FRUIT CAKES

Rich fruit cakes may be stored in the freezer, but if space is limited it is better to store them in tins. Dundee cakes, sultana cakes and other light fruit mixtures freeze very well. They should be thawed in wrappings at room temperature.

SMALL CAKES

Small fruit cakes and sponge drops can be frozen in bags or boxes; in boxes, the layers of cakes should be divided by Cellophane or greaseproof paper. Cupcakes may be made in paper and foil cases, iced and frozen, then packed in boxes with paper between the layers. Cakes to be cut in squares can be frozen in the baking tin, or in a baking conatiner made of heavy duty foil, which can be covered with foil, or put into a polythene bag for freezing, and then the cake may be cut in squares when thawed.

Choux pastry and meringues can be frozen successfully if unfilled. They are best frozen in single layers then packed in boxes because of their delicacy.

BISCUITS

Baked biscuits freeze well, but store just as well in tins, so that freezer space can be saved. Uncooked frozen biscuits are extremely useful, and the frozen dough will give light crisp biscuits. The best way to prepare biscuits is to freeze batches of any recipe in cylinder shapes in polythene or foil, storing them carefully to avoid dents in the freezer from other packages.

This unfrozen dough should be left in its wrappings in the refrigerator for 45 minutes until it begins to soften, then cut in slices and baked. If the dough is too soft, it will be difficult to cut. Unbaked biscuit dough will store for 2 months.

Baked biscuits should be packed in layers in cartons with Cellophane or greaseproof paper between layers, and with crumpled up paper in air spaces to safeguard freshness and stop breakages.

VERY LIGHT SPONGECAKE
3 eggs
5 oz. caster sugar
3 oz. plain flour

Separates yolks and whites of eggs and whisk whites until very stiff. Put the mixing bowl over a saucepan of hot, but not boiling, water and gradually beat in yolks and sugar. Beat for 5 minutes, then fold in sifted flour very carefully. Bake in an 8 ins. greased tin with greaseproof paper on the bottom at 400°F (Gas Mark 6) for 45 minutes. Cool thoroughly.
Pack in polythene bag or heavy duty foil.
To serve thaw in wrappings at room temperature for 3 hours.
Storage Time 10 months.

VICTORIA SANDWICH
4 oz. margarine
4 oz. caster sugar
2 eggs
4 oz. self-raising flour

Cream margarine and sugar until light and fluffy and almost white. Break in eggs one at a time and beat well. Fold in sifted flour. Bake in two 7 ins. greased tins at 335°F (Gas Mark 3) for 30 minutes. Leave in tins for 2 minutes, then cool on wire rack.
Pack in polythene bag or heavy duty foil with a piece of Cellophane or greaseproof paper between the two layers.
To serve thaw in wrappings at room temperature for 3 hours, fill and ice as required.
Storage Time 4 months.

SPONGE DROPS
2 eggs
3 oz. caster sugar
3 oz. plain flour
¼ teaspoon baking powder
Pinch of salt

Separate yolks and whites. Add salt to whites and whisk until very stiff. Gradually whisk in sugar and yolks alternately until the mixture is thick and creamy. Fold in flour sifted with baking powder. Put in spoonsful on greased and floured baking

sheets. Dust with sugar and bake at 450°F (Gas Mark 8) for 5 minutes. Cool thoroughly.
Pack in polythene bags.
To serve thaw in wrappings at room temperature for 1 hour, then sandwich together with jam, or with jam and whipped cream, and dust with icing sugar.
Storage Time 10 months.

DUNDEE CAKE

- 8 oz. butter
- 8 oz. caster sugar
- 5 eggs
- 8 oz. self-raising flour
- ½ teaspoon ground nutmeg
- 12 oz. mixed currants and sultanas
- 3 oz. ground almonds
- 3 oz. chopped glacé cherries
- 2 oz. chopped candied peel
- 2 oz. split blanched almonds

Cream butter and sugar until fluffy and add eggs one at a time with a sprinkling of flour to stop curdling. Beat well after adding each egg. Stir in flour, ground almonds and the fruit lightly coated with a little of the flour. Put into greased and lined 10 ins. tin, smooth top and arrange almonds on top. Bake at 325°F (Gas Mark 3) for 2½ hours. Cool thoroughly.
Pack in polythene bag or heavy duty foil.
To serve thaw in wrapping at room temperature for 3 hours.
Storage Time 4 months.

LUNCHEON CAKE

- 4 oz. butter
- 8 oz. caster sugar
- 3 eggs
- 6 oz. plain flour
- ½ teaspoon baking powder
- ¼ teaspoon salt
- ½ teaspoon ground nutmeg
- 2 tablespoons milk
- 2 tablespoons honey
- ¼ teaspoon bicarbonate of soda
- 8 oz. walnuts
- 1 lb. seedless raisins

Cream butter and sugar until light and fluffy. Beat eggs together and add to creamed mixture with sifted flour, baking powder, salt and nutmeg. Stir in milk. Mix honey and bicarbonate of soda together and add to mixture, and stir in walnuts and raisins. Put mixture into greased and floured 2 lb. loaf tin.

Bake at 325°F (Gas Mark 3) for 2¼ hours. Cool in tin before turning out.
Pack in polythene bag or foil.
To serve thaw in wrappings at room temperature for 3 hours.
Storage Time 4 months.

SUGAR CAKE

8 tablespoons water	6 oz. plain flour
2 oz. butter	2 teaspoons baking powder
3 eggs	1 tablespoon grated lemon rind
8 oz. caster sugar	

Bring water to boiling point, add butter and cool. Beat eggs and sugar until white and fluffy, and add flour, baking powder, lemon and water mixture. Stir until well blended, and pour into 8 ins. cake tin which has been buttered and dusted with fine breadcrumbs. Bake at 325°F (Gas Mark 3) for 1 hour. Cool on wire rack.
Pack in polythene bag or foil.
To serve thaw in wrappings at room temperature for 3 hours. Dust with sugar.
Storage Time 4 months

ORANGE LOAF

2 oz. butter	2 tablespoons milk
6 oz. caster sugar	7 oz. plain flour
1 egg	2½ teaspoons baking powder
Grated rind and juice of 1 small orange	½ teaspoon salt

Cream butter and sugar until fluffy, and gradually add beaten egg with orange rind and juice, and milk. Add flour sieved with baking powder and salt, and fold into creamed mixture. Put into greased 2 lb. loaf tin and bake at 375°F (Gas Mark 5) for 1 hour. Cool on wire rack.
Pack in polythene bag or in foil.
To serve thaw in wrappings at room temperature for 3 hours, then slice and spread with butter.
Storage Time 4 months.

GINGERBREAD

8 oz. golden syrup
2 oz. butter
2 oz. sugar
1 egg
8 oz. plain flour
1 teaspoon ground ginger
1 oz. candied peel
1 teaspoon bicarbonate of soda
Milk

Melt syrup over low heat with butter and sugar, and gradually add to sifted flour and ginger together with beaten egg. Mix soda with a little milk and beat into the mixture, and add chopped peel. Pour into rectangular tin and bake at 325°F (Gas Mark 3) for 1 hour. Cool in tin.

Pack baking tin into polythene bag or foil; or cut gingerbread in squares and pack in polythene bags or boxes.

To serve Thaw in wrappings at room temperature for 2 hours. This cake may be served as a pudding with apple purée and cream or ice cream.

Storage Time 2 months.

CHOCOLATE CAKE (or COFFEE CAKE)

4 oz. margarine
5 oz. caster sugar
4 oz. self-raising flour
1 tablespoon cocoa or 1 tablespoon coffee essence
2 eggs
1 tablespoon milk

Slightly soften margarine, put all ingredients into bowl and blend until creamy and smooth. Bake in two 7 ins. tins at 350°F (Gas Mark 4) for 30 minutes. Cool and fill with *icing*. Make this by blending together 6 oz. icing sugar, 1 oz. cocoa (or 1 tablespoon coffee essence), 2 oz. soft margarine and 2 dessertspoons hot water. Fill and put layers together, and put more icing on top of cake.

Pack in polythene bag or foil. It is easier to freeze the cake without wrapping, then pack for storage.

To serve remove from wrappings and thaw at room temperature for 3 hours.

Storage Time 4 months.

GOLDEN LEMON CAKE

1½ oz. butter	6 oz. self-raising flour
6 oz. caster sugar	¼ pint milk
3 egg yolks	Pinch of salt
¼ teaspoon lemon essence	

Cream butter and sugar until fluffy and slowly add egg yolks and lemon essence. Add flour alternately with milk and the pinch of salt. Bake in two 8 ins. tins lined with paper at 350°F (Gas Mark 4) for 25 minutes. Cool, fill and ice with *Lemon Frosting*. Make this by blending together 3 tablespoons butter, 1 tablespoon grated orange rind, 2 tablespoons lemon juice, 1 tablespoon water and 1 lb. icing sugar.

Pack in polythene bag or foil. It is easier to freeze the cake without wrapping, then pack for storage.

To serve remove from wrappings and thaw at room temperature for 3 hours.

Storage Time 4 months.

CHOUX PASTRY

¼ pint water	Pinch of salt
2 oz. lard	2 small eggs
2¼ oz. plain flour	

Bring water and lard to boil in pan and immediately put in flour and salt. Draw pan from heat and beat until smooth with wooden spoon. Cook for 3 minutes, beating very thoroughly, cool slightly and beat in whisked eggs until the mixture is soft and firm but holds its shape. Pipe in finger lengths on baking sheets and bake at 425°F (Gas Mark 7) for 30 minutes. For *Cream Puffs*, put mixture in small heaps, cover baking tin with a roasting tin, and bake at 425°F (Gas Mark 7) for 1 hour. Cool. *Pack* in polythene bags.

To serve thaw in wrappings at room temperature for 2 hours, fill with whipped cream, and top with chocolate or coffee icing.

Storage Time 1 month.

BROWNIES

8 oz. granulated sugar	2 eggs
1½ oz. cocoa	2 tablespoons creamy milk
3 oz. self-raising flour	4 oz. melted butter or margarine
½ teaspoon salt	

Optional – 3 oz. shelled walnuts *or* seedless raisins.

Stir together sugar, cocoa, flour and salt. Beat eggs and milk, and add to the dry mixture, together with melted fat. Stir in broken walnuts or raisins. Pour into rectangular tin (about 8 x 12 ins.) and bake at 350°F (Gas Mark 4) for 30 minutes. Cool in tin.

Pack by covering baking tin with foil, or by putting tin into polythene bag. A baking tin may be made of heavy duty foil for cooking and freezing if a normal baking tin cannot be spared for storage.

To serve thaw in wrappings at room temperature for 3 hours, then top with 4 oz. melted plain chocolate.

Storage Time 4 months.

CHOCOLATE CRUMB CAKE

4 oz. butter
1 tablespoon sugar
2 tablespoons cocoa
1 tablespoon golden syrup
8 oz. fine biscuit crumbs

Cream butter and sugar and add cocoa and syrup. Mix well and blend in biscuit crumbs. Press mixture into greased foil tray about 1 ins. deep.

Pack by covering foil tray with foil lid.

To serve Remove foil lid and thaw at room temperature for 3 hours, then top with 2 oz. melted plain chocolate, leave to set, and cut in small squares.

Storage Time 4 months.

BASIC SUGAR BISCUITS

4 oz. butter
8 oz. caster sugar
1 egg or 2 egg yolks
1 tablespoon milk
½ teaspoon vanilla essence
6 oz. plain flour
½ level teaspoon baking powder
½ level teaspoon salt

Soften butter and work in sugar, egg, milk and vanilla. Add sifted flour, baking powder and salt and work into a firm dough. Chill mixture, then form into cylinder shape about 2 ins. diameter.

VARIATIONS

Butterscotch Biscuits – Use brown sugar and 1 oz. chopped nuts.

Chocolate Biscuits – Add 1 oz. cocoa.

Date Biscuits – Add 2 oz. chopped dates.

Ginger Sugar Biscuits – Add 1 teaspoon ground ginger.

Lemon Biscuits – Add ½ teaspoon lemon essence instead of vanilla.

Nut Biscuits – Add 2 oz. chopped nuts.

Orange Biscuits – Use orange juice instead of milk, and add the grated rind of ½ orange.

Pack cylinder of dough in heavy duty foil or polythene bag.

To serve thaw in wrappings in refrigerator for 45 minutes, cut in slices and place on baking sheet. Bake at 375°F (Gas Mark 5) for 10 minutes. Cool on wire rack.

Storage Time 2 months.

Chapter Ten

YEAST BREADS AND BUNS, TEA BREADS, SCONES, PANCAKES

Bread, buns, scones and pancakes freeze extremely well, and even items like crumpets, muffins and doughnuts come out of the freezer as fresh as they went in. In households where packed meals are prepared, or there are many children and long school holidays, bulk supplies of these foods are very useful. Home baking becomes a worthwhile proposition when large batches of yeast-baked bread and buns can be frozen, or scones and pancakes prepared in double or treble quantities to provide quick meals.

INGREDIENTS
Fresh yeast can be stored in the freezer if supplies are erratic. The yeast can be bought in 8 oz. or 1 lb. packages, divided into 1 oz. cubes and each cube wrapped in polythene or foil, and then a quantity of these stored in a preserving jar in the freezer. A cube of yeast will be ready for use after 30 minutes at room temperature.

Bread flour gives the best results with yeast; this flour made from hard wheat is now available under the name of "Country Life" in 3 lb. bags. This flour is also best for flaky pastry and for yeast pastries.

UNCOOKED YEAST MIXTURES
Unbaked bread and buns may be frozen for up to 2 weeks, but proving after freezing takes a long time, and the texture may be heavier. Unbaked dough to be frozen should be allowed to prove once, then be shaped for baking, or kept in bulk if this is easier for storage. The surface should be brushed with a little olive oil or unsalted melted butter to prevent toughening of the crust, and a little extra sugar added to sweet mixtures.

Single loaves or a quantity of dough can be packed in foil or polythene; rolls can be packed in layers separated by Cellophane before wrapping in foil or polythene.

The dough should be thawed in a moist warm place; greater speed in thawing will give a lighter texture loaf. After shaping, the mixture must be proved again before baking. If the bread has been shaped before freezing, it should be proved once in a warm place before baking.

COOKED YEAST MIXTURES

Cooked yeast mixtures such as bread, rolls and buns, freeze particularly well when one day old. They can be packed in foil or in polythene bags in the required quantities, and should be thawed in wrappings at room temperature. $1\frac{1}{2}$ lbs. loaf will take 3 hours to thaw. Bread may also be thawed in a moderate oven, and will be like freshly baked bread, but will become stale very quickly if this method is used.

RICH YEAST MIXTURES

Yeast doughs enriched with eggs and sugar, such as used for tea breads, savarins and babas, will keep for 3 months in the freezer. Yeast pastries, such as Danish pastries, which have fat incorporated, are best kept for only 1 month.

SCONES

Scones made with baking powder, or with bicarbonate of soda and cream of tartar, stale quickly, but large batches can be quickly made and packaged in suitable quantities for the freezer.

Unbaked scones can be frozen and kept for 2 weeks; they may be partly thawed and then cooked, or put straight into a hot oven without thawing.

Baked scones will keep for 2 months, and for ease of packing and service, it is preferable to freeze them in their finished state rather than unbaked. They are most easily packaged in required quantities in polythene bags (they may be wasted if packed in too large quantities).

Scones should be thawed in wrappings at room temperature for 1 hour, or at 350°F (Gas Mark 4) for 10 minutes with a covering of aluminium foil. Frozen scones may also be split and put under a hot grill.

PANCAKES, GRIDDLECAKES AND DROP SCONES

Pancakes, griddlecakes and drop scones should be cooked as usual and cooled before packing. Large thin pancakes should be packed with a layer of Cellophane or greaseproof paper between each, stacked like a cake, then wrapped in foil or polythene for the freezer. They may easily be separated while frozen, or thawed in one piece in wrappings at room temperature; they can then be wrapped round a sweet or savoury filling and heated in a low oven or on a plate over steam, covered with a cloth. Griddlecakes and drop scones should be thawed in wrappings at room temperature before serving.

WHITE BREAD

2½ lbs. white bread flour
1 oz. fresh yeast
2 oz. fat (butter, lard or margarine)
½ oz. salt
1½ pints warm water

Warm a large bowl and put in flour. Make a well in centre and sprinkle salt round edge. Cream yeast with a little warm water and pour into the well. Add remaining water and warmed fat and mix well to a consistency like putty. Leave to prove until double its size. Divide into loaf tins and leave to prove again until bread reaches top of tins. Bake at 450°F (Gas Mark 8) for 45 minutes, turning bread once in the oven. Cool on a rack and leave overnight.
Pack in polythene bags or in heavy duty foil.
To serve thaw in wrappings at room temperature for 3 hours.
Storage Time 8–12 months.

CURRANT BREAD

1½ lbs. white bread flour
4 oz. sugar
Pinch of salt
1 oz. fresh yeast
4 oz. warm butter
½ pint warm milk
8 oz. mixed dried fruit
2 oz. chopped mixed peel

Using a large warm bowl, mix flour, sugar and salt, and add yeast creamed with a little sugar. Work in butter and milk, knead well and leave to prove for 1½ hours. Work in fruit and peel, and put

into loaf tins or shape into buns. Prove for 45 minutes. Bake at 375°F (Gas Mark 5) for 45 minutes, turning loaves after 20 minutes; small buns should be baked for 20 minutes. Cool on a wire tray, brushing with a mixture of milk and sugar to give a sticky finish. Store overnight in a tin.
Pack in polythene bags or in heavy duty foil.
To serve thaw in wrappings at room temperature for 3 hours (loaves), 1½ hours (buns).
Storage Time 2 months.

BAPS

1 lb. white bread flour	2 level teaspoons salt
2 oz. lard	1 oz. yeast
1 level teaspoon sugar	½ pint lukewarm milk and water

Sieve the flour and rub in the lard and sugar. Dissolve the salt in half the liquid, and cream the yeast into the rest of the liquid. Mix into the flour, knead and prove until double in size. Divide into pieces and make into small flat rounds about 4 ins. across. Brush with milk, put on a greased baking sheet, prove again, and bake at 450°F (Gas Mark 8) for 20 minutes. Cool on a rack.
Pack in polythene bags. Baps may be split, buttered and fillled before freezing (see SANDWICHES AND FILLINGS).
To serve thaw in wrappings at room temperature for 1 hour.
Storage Time Unfilled: 10–12 months. Filled: 1 month.

BRIDGE ROLLS

2½ lbs. white bread flour	½ oz. salt
1 oz. fresh yeast	2 eggs
4 oz. butter or margarine	1 pint warm milk

Make the bread dough as for White Bread, whisking the eggs into the milk before adding to the flour. Shape rolls into small sausage shapes and prove for 30 minutes. Paint with milk, and bake at 450°F (Gas Mark 8) for 15 minutes. Cool on rack.
Pack in polythene bags in required quantities.
To serve thaw in wrappings at room temperature for 45 minutes.
Storage Time 10–12 months.

CROISSANTS

1 oz. butter
¼ pint warm milk
1 heaped teaspoon salt
1½ tablespoons sugar
1 oz. fresh yeast dissolved in a little warm water

12 oz. white bread flour
4 oz. butter
1 egg yolk beaten with a little milk

Put butter in bowl, pour on warm milk and add salt and sugar. Cool to lukewarm, then add yeast and gradually add flour to give a soft dough. Cover dough with damp cloth and leave for 2 hours. Knead dough, chill thoroughly, and roll into a rectangle. Spread butter lightly and evenly over dough. Fold over dough to a rectangle and roll again. Chill, roll and fold twice more at intervals of 30 minutes. Roll dough out to ¼ ins. thickness and cut into 4 ins. squares. Divide each square into 2 triangles, and roll each triangle up, starting at longest edge and rolling towards the point. Bend into crescent shapes, put on floured baking sheet, brush with beaten egg and milk, and bake at 425°F (Gas Mark 7) for 15 minutes. Cool.

Pack into polythene bags. Store carefully to avoid crushing and flaking.

To serve thaw in wrappings at room temperature for 30 minutes, then lightly heat in the oven or under the grill.

Storage Time 2 months.

BRIOCHE

8 oz. white bread flour
1 oz. yeast
2 tablespoons warm water
3 eggs

6 oz. melted butter
1 teaspoon salt
1 tablespoon sugar

Put 2 oz. flour in a warm bowl and mix with yeast creamed with a little warm water. Put the little ball of dough into a bowl of warm water and it will expand and form a sponge. Put remaining flour into a bowl and beat in eggs thoroughly. Add butter, salt and sugar and continue beating. Add yeast sponge removed from water, and mix well. Cover with damp cloth and prove for 2 hours, then knead dough and leave in cool place overnight. Half-fill castle pudding tins with dough and prove 30

minutes. Bake at 450°F (Gas Mark 8) for 15 minutes. Cool.
Pack in polythene bags.
To serve thaw in wrappings at room temperature for 45 minutes. Brioche may also be heated with the tops cut off and the centres filled with sweet or savoury mixtures (e.g. creamed mushrooms, chicken or shrimps).
Storage Time 2 months.

MUFFINS

1 egg	1 lb. white bread flour
½ pint milk	1 teaspoon salt
1 oz. butter or margarine	½ oz. fresh yeast

Beat together egg, milk and warm fat. Put flour and salt into a bowl and pour in yeast creamed with a little warm water. Add butter, milk and egg mixture, and knead thoroughly to a soft but not sticky dough. Cover and prove for 1½ hours. Roll out dough to ½ ins. thickness on a floured board and cut out muffins with a large tumbler. Bake on a griddle, turning as soon as the bottoms are browned, or on a baking sheet at 450°F (Gas Mark 8) for 15 minutes, turning half way through cooking. Cool.
Pack in polythene bags.
To serve thaw in wrappings at room temperature for 30 minutes, split and toast.
Storage Time 10–12 months.

TEA CAKES

8 oz. white bread flour	1 teaspoon sugar
Pinch of salt	1 egg
½ oz. butter	¼ pint milk
½ oz. fresh yeast	2 oz. sultanas

Warm flour and salt and rub in butter. Cream yeast and sugar and mix into the flour together with the egg, milk and sultanas. Knead well and leave in a warm place until double in size. Knead again and divide into three, shaping into round flat cakes. Put on greased tins and prove for 10 minutes. Bake at 450°F (Gas Mark 8) for 12 minutes. Brush over with milk and sugar immediately after removing from the oven. Cool.
Pack in polythene bags

To serve thaw in wrappings at room temperature for 45 minutes.
Storage Time 2 months.

DOUGHNUTS

8 oz. white bread flour	¼ pint lukewarm milk
1 teaspoon salt	½ oz. margarine
2 teaspoons sugar	Raspberry jam
½ oz. yeast	

Sieve flour and salt, and mix in sugar. Whisk yeast in half the milk; add melted margarine to remaining milk and cool to lukewarm. Mix both liquids into flour and knead well, then leave to prove for 1 hour. Knead lightly, divide into 16 pieces and form into balls. Flatten balls, put jam in centre, and fold edges to enclose jam, pressing together firmly. Prove for 20 minutes. Fry until golden in hot fat and drain very thoroughly on kitchen paper.
Pack in polythene bags.
To serve heat straight from freezer at 400°F (Gas Mark 6) for 8 minutes, then roll in caster sugar and serve at once.
Storage Time 1 month.

DANISH PASTRIES

8 oz. white bread flour	½ oz. yeast
¼ level teaspoon salt	¼ pint warm water
2½ oz. sugar	3 oz. butter

Put flour and salt in a warm basin. Cream yeast with a little of the sugar and put into the flour together with the remaining sugar and water. Mix to a soft, slightly sticky dough, and leave to rise in a warm place until increased by one-third in volume. Form butter into a rectangle and dust with flour. Flatten dough with hands and fold with the fat in the centre like a parcel. Roll and fold twice like puff pastry. Leave in a cold place for 20 minutes, then roll and fold twice more and leave for 20 minutes. Roll out to ¼ ins. thick. Fold squares of pastry over fillings of dried fruit, jam, marzipan or chopped nuts, to make envelope shapes. Brush pastries with a mixture of melted butter, milk and egg, and bake without proving at 375°F (Gas Mark 5) for

30 minutes. Cool. The pastries may be frozen uniced or with a light water icing.
Pack in foil trays with foil lid, or put trays into polythene bags.
To serve thaw at room temperature, removing wrappings if iced, for 1 hour.
Storage Time 2 months

BASIC SCONES

1 lb. plain white flour	3 oz. butter
1 teaspoon bicarbonate of soda	¼ pint milk
2 teaspoons cream of tartar	

Sift together flour, soda and cream of tartar and rub in butter until mixture is like breadcrumbs. Mix with milk to soft dough. Roll out, cut in rounds, and place close together on greased baking sheet. Bake at 450°F (Gas Mark 8) for 12 minutes. Cool.

Fruit Scones
Add 1½ oz. sugar and 2 oz. dried fruit.
Cheese Scones
Add pinch each of salt and pepper and 3 o.z grated cheese.
Pack in sixes or dozens in polythene bags.
To serve thaw in wrappings at room temperature for 1 hour, or heat at 350°F (Gas Mark 4) for 10 minutes with a covering of foil.
Storage Time 2 months.

DROP SCONES

8 oz. plain white flour	1 level teaspoon cream of tartar
¼ level teaspoon salt	1 level tablespoon sugar
½ level teaspoon bicarbonate of soda	1 egg
	¼ pint milk

Sieve together flour, salt, soda and cream of tartar. Stir in sugar and mix to a batter with egg and milk. Cook in spoonsful on lightly greased griddle or frying pan. When bubbles appear on the surface, turn quickly and cook other side. Cool in a cloth to keep soft.
Pack in foil, with a sheet of Cellophane or greaseproof paper between layers.

To serve thaw in wrappings at room temperature.
Storage Time 2 months.

WELSH GRIDDLECAKES

1 lb. plain flour	2 oz. currants
Pinch of salt	Pinch of nutmeg
½ teaspoon baking powder	1 egg
3 oz. butter	Water
3 oz. sugar	

Sieve together flour, salt and baking powder and rub in butter until the mixture is like breadcrumbs. Add sugar, currants, nutmeg and egg, and a little water to make a stiff dough. Roll out and cut into rounds. Bake on both sides until golden on a hot greased griddle or thick frying pan. Cool.
Pack in polythene bags.
To serve put on baking tray with covering of foil and heat at 350°F (Gas Mark 4) for 10 minutes, and eat with butter.
Storage time 2 months.

GRIDDLE BREAD

8 oz. wholemeal flour	1 level teaspoon salt
8 oz. plain flour	1 dessertspoon dripping
1 level dessertspoon sugar	Milk
1 level teaspoon bicarbonate of soda	

Mix wholemeal and plain flour, and add sugar, soda and salt. Rub in dripping and mix to a dough with the milk which is stiff but will roll easily. Roll in a round 1 ins. thick and cut into four sections. Cook on a hot griddle or frying pan for 10 minutes each side. Cool.
Pack in polythene bags.
To serve thaw in wrappings at room temperature for 1 hour.
Storage Time 2 months

BASIC PANCAKES

4 oz. plain flour	½ pint milk
¼ teaspoon salt	1 tablespoon oil or melted butter
1 egg and 1 egg yolk	

Sift flour and salt and mix in egg and egg yolk and a little milk. Work together and gradually add remaining milk, beating to a smooth batter. Fold in oil or melted butter. Fry large or small thin pancakes.

Pack in layers separated by Cellophane and put in foil or polythene bag.

To serve separate the pancakes, put on a baking sheet and cover with foil, and heat at 400°F (Gas Mark 6) for 10 minutes.

Storage Time 2 months.

Chapter Eleven

SANDWICHES AND FILLINGS

Bread freezes extremely well, so that sandwiches may be frozen in advance for packed meals or for parties. Keeping time will depend on individual fillings, but there is little point in storing sandwiches for longer than 4 weeks.

TYPES OF BREAD

As well as plain white bread, whole wheat, rye, pumpernickel and fruit breads can be used for frozen sandwiches, together with baps and rolls. Brown bread is good for fish fillings, and fruit bread for cheese or sweet fillings.

FILLINGS

A tremendous variety of fillings may be used, but one or two items should not be included. Cooked egg whites become tough and dry in the freezer; raw vegetables such as celery, lettuce, tomatoes and carrots are not successful; salad cream and mayonnaise curdle and separate when frozen and will soak into the bread when thawed; jam also soaks into the bread during thawing.

PACKING AND FREEZING

Sandwiches are best packed in groups of six or eight rather than individually. An extra slice or crust of bread at each end of the package will prevent drying out. Heavy duty foil or polythene sheeting or bags can be used for packaging. If sandwiches are frozen against the freezer wall, this will result in uneven thawing. The packages should be frozen a few inches from the wall of the freezer, with the edges towards the wall. They should be thawed in their wrappings in the refrigerator for 12 hours, or at room temperature for 4 hours.

PREPARING SANDWICHES FOR FREEZING

When preparing a quantity of sandwiches for freezing, it is a good idea to use an assembly-line technique to speed up production.

1. Soften butter or margarine, but do not melt.

2. Prepare fillings and refrigerate ready for use.
3. Assemble wrapping materials.
4. Assemble breads and cut (or split rolls or baps).
5. Spread bread slices with fat, going right to the edge to prevent fillings soaking in.
6. Spread fillings evenly on bread to ensure even thawing.
7. Close and stack sandwiches.
8. Cut with a sharp knife. Leave in large portions, such as half-slices, and leave crusts on, which will help to prevent sandwiches being misshapen during freezing. Sandwiches may be cut smaller and the edges trimmed before serving.
9. Wrap sandwiches tightly in Cellophane, then in foil or polythene. With an inner wrapping, the other covering may be removed and kept for further use, and the inner package taken in a lunch box for thawing.
10. Label and freeze.

PARTY SANDWICHES

Special types of sandwiches may be usefully frozen for such events as cocktail parties or weddings. Sandwiches which are rolled, or formed into pinwheels or ribbons are best frozen in aluminium foil trays to keep their shape, covered with foil and carefully sealed before freezing. Thaw for 12 hours in the refrigerator, or four hours at room temperature; thawing may be speeded up if the foil is replaced by waxed paper when the tray is removed from the freezer.

ROLLED SANDWICHES

Use finely grained bread and well creamed butter, and cut bread very thinly. It will assist rolling, if the bread is lightly rolled with a rolling pin before spreading. Spread with butter and filling. Roll sandwiches and pack closely in tray. A creamed filling may be used for these sandwiches, or the bread rolled round tinned asparagus tips, or lightly cooked fresh ones.

PINWHEEL SANDWICHES

Use a sandwich loaf and cut slices lengthwise. Spread with creamed butter and filling, and roll up bread like a Swiss roll. Pack in foil trays. For serving, thaw and cut in slices of required thickness.

RIBBON SANDWICHES
For each sandwich, use three slices of bread, alternating white and brown. Spread with butter and filling and make a triple sandwich. Press lightly under a weight before packing and freezing. Thaw and cut in finger-thick slices for serving.

FILLINGS
Butter or margarine may be used for spreading sandwiches. These may be given additional flavouring according to the sandwich filling. The fat should be slightly softened, but not melted, and flavoured with lemon juice, grated horseradish, grated cheese, chopped parsley, tomato puree. Peanut butter can also be used.

CHEESE
Cream cheese with olives and peanuts.
Cream cheese with chutney.
Cream cheese with chopped dates, figs or prunes.
Cottage cheese with orange marmalade or apricot jam.
Blue cheese with roast beef.
Blue cheese with chopped cooked bacon.
Cheddar cheese and chopped olives or chutney.
Cream cheese with liver sausage.

FISH
Mashed sardines, hard-boiled egg yolk and a squeeze of lemon juice.
Minced shrimps, crab or lobster with cream cheese and lemon juice.
Tuna fish with chutney.
Mashed canned salmon with cream cheese and lemon juice.

MEAT AND POULTRY
Sliced tongue, corned beef, luncheon meat or ham with chutney.
Sliced roast beef with horseradish.
Sliced roast lamb with mint jelly.
Sliced chicken or turkey with ham and chutney.
Sliced duck or pork with apple sauce.
Minced ham with chopped pickled cucumber and cream cheese.

Chapter Twelve

SCHOOL HOLIDAY FOOD

These dishes are not designed solely for eating in the school holidays. They are the basic items around which family meals may be planned at any time; recipes which are equally suitable for mid-day on a busy Saturday, for an evening meal on a quiet Sunday.

All these dishes are inexpensive to prepare, and can be made in sufficiently large quantities to cope with unexpected numbers at holiday time. To supplement these dishes, bulk stocks of peas, carrots and chips will be most useful. Commercially prepared beefburgers, fish fingers and ice cream are popular with children, and can be very useful for the rapid preparation of high tea or light supper. The freezer can also be stocked with baps for making hamburgers, buns and doughnuts to fill in the corners.

Suitable recipes for holiday periods will also be found in STARTERS: MAIN COURSES: SWEET COURSES: PIES AND FILLINGS: and PASTA.

FISH CAKES

1 lb. cooked white fish	2 oz. butter
1 lb. mashed potato	Salt and pepper
4 teaspoons chopped parsley	2 small eggs

Mix flaked fish, potato, parsley, melted butter and seasonings together, and bind with egg. Divide the mixture into sixteen pieces and form into flat rounds. Coat with egg and breadcrumbs and fry until golden. Cool quickly.
Pack in polythene bags or waxed cartons, separating fishcakes with waxed paper or Cellophane. Fishcakes may also be frozen uncovered on baking sheets, and packed when solid.
To serve reheat in oven or frying pan with a little fat, allowing 5 minutes' cooking on each side.
Storage Time 1 month.

FISH PIE

1 lb. cooked white fish	2 tablespoons chopped parsley
1 lb. cooked potato	Salt and pepper
½ pint milk	½ pint white sauce
2 oz. butter	

Flake the fish. Mash the potato with warmed milk, butter, parsley and seasonings. Mix the fish with the sauce, and put in base of greased pie dish or foil container. Spread potato mixture on top.

Pack by putting container into Polythene bag, or covering with a lid of heavy duty foil. Freeze until firm.

To serve put dish in cold oven and cook at 400°F (Gas Mark 6) for 1½ hours. The dish may be thawed for 3 hours in a refrigerator and cooked at 425°F (Gas Mark 7) for 25 minutes.

Storage Time 1 month.

KEDGEREE

1 lb. cooked smoked haddock or smoked cod	Salt and pepper
	1 tablespoon chopped parsley
8 oz. Patna rice	2 hardboiled egg yolks
3 oz. butter	

Optional – 4 oz. button mushrooms and/or 1 green pepper

Flake the fish. Cook rice in fast-boiling salted water for about 20 minutes until tender, and drain well. Mix with flaked fish, melted butter, seasonings, parsley and chopped egg yolks. For a supper dish, add chopped button mushrooms and/or green pepper cooked in a little butter until soft.

Pack in foil container covered with lid of heavy duty foil, or in Polythene bag.

To serve thaw in refrigerator for 3 hours, then reheat in double saucepan over boiling water. The dish may be put into a cold oven straight from the freezer and heated at 400°F (Gas Mark 6) for 45 minutes. Heating will be speeded if the dish is stirred occasionally.

Storage Time 1 month.

MEAT LOAF

2 eggs	1½ teaspoons salt
½ pint milk	½ teaspoon pepper
6 oz. soft white breadcrumbs	2 lbs. minced chuck steak

Optional – Concentrated tomato purée and/or mixed herbs.

Beat eggs lightly, then work in milk, breadcrumbs, seasonings and mince. Add additional flavour if liked with tomato purée and/or mixed herbs. Mix well and put into a baking tin lined with foil, overlapping 6 ins. above the top of the pan. Use a pan 9 inches x 9 inches x 2 inches, or use two small or one large loaf tins. The meat loaf may be frozen uncooked or cooked.

Pack uncooked meat loaf by folding over foil to form parcel, freezing and then removing foil parcel from tin for storage. Alternatively, the meat loaf may be cooked at 350°F (Gas Mark 4) for 1 hour 40 minutes, cooled, parcelled and frozen.

To serve uncooked meat loaf, remove foil from meat, put meat into pan and bake at 350° (Gas Mark 4) for 1 hour 40 minutes. The frozen cooked meat loaf should be heated at 400°F (Gas Mark 6) for 45 minutes, or it may be thawed in the refrigerator and used cold in sandwiches.

Storage Time 1 month.

COTTAGE PIE

1 lb. cooked beef, or lamb
¼ pint stock
Salt and pepper
1 medium onion
1 lb. mashed potato

This recipe may be made with fresh minced meat, but in most households is prepared from left over roast meat. Mince the meat and moisten with stock, seasoning to taste. Mix with the onion which has been chopped and lightly cooked in a little fat. Put into foil container and cover with mashed potato. Reconstituted mashed potato powder can be used with good results.

Pack foil container in polythene bag, or cover with a lid of heavy duty aluminium foil.

To serve put in cold oven and reheat at 400°F (Gas Mark 6) for 45 minutes.

Storage Time 1 month.

STUFFED MARROW RINGS

1 medium marrow
1 lb. cooked beef or lamb
6 oz. fresh white breadcrumbs
1 medium onion
½ pint stock
Salt and pepper

Cut marrow into 2 inch slices, removing seed and pith, and cook in boiling water for 3 minutes. Drain well and arrange in foil container or greased oven dish. Mince the meat and mix with breadcrumbs, the onion which has been chopped and softened in a little fat, stock, and seasonings. A little tomato purée may be added for flavouring, and the stock may be thickened with a little cornflour if a firmer mixture is liked. Cook the mixture together for 10 minutes, then fill marrow rings.
Pack by covering dish with lid of heavy duty foil.
To serve reheat at 375°F (Gas Mark 5) for 1½ hours, removing lid for final 15 minutes.
Storage Time 1 month.

BAKED STUFFED POTATOES
Large potatoes			Butter
Milk				Salt and pepper

Scrub potatoes, prick with a fork, and bake at 350°F (Gas Mark 4) for 1½ hours. Scoop pulp from shells, mash with milk, butter and seasonings and return to shells. Potatoes may also be stuffed with a little creamed smoked fish, creamed kidneys, chicken or ham.
Pack each potato in a piece of heavy duty foil.
To serve reheat at 350°F (Gas Mark 4) for 40 minutes. A little grated cheese may be sprinkled on the potatoes for the last 10 minutes, opening the foil so that the cheese will brown slightly.
Storage Time 3 months (1 month if fish or meat fillings are used).

BAKED APPLES
Large apples			Spice
Brown sugar			Lemon juice

Use large firm fruit and wash apples well. Remove cores, leaving ¼ ins. at bottom to hold filling. Fill with brown sugar, a pinch of powdered cloves or cinnamon and a squeeze of lemon juice. Bake at 400°F (Gas Mark 6) until apples are tender.
Pack in individual waxed tubs or foil dishes. Quantities of apples in a large container may be separated by Cellophane. Cover with

lid of heavy foil if individual waxed tubs are not used.
To serve reheat under foil lid at 350°F (Gas Mark 4) for 30 minutes. These apples may be eaten cold.
Storage Time 1 month (longer if spice is omitted).

FRUIT FRITTERS

4 oz. plain flour
Pinch of salt
1 egg and 1 egg yolk
½ pint milk
1 tablespoon melted butter
1 tablespoon fresh white breadcrumbs
6 eating apples or 6 bananas or 1 large can pineapple rings

Prepare batter by mixing together flour, salt, egg and egg yolk and milk, and folding in melted butter and breadcrumbs. Peel, core and slice apples into ¼ ins. rings, or cut bananas in half lengthways. Dip fruit into batter and fry until golden in deep fat. Drain on absorbent paper and cool.
Pack in polythene bags, foil or waxed containers, separating fritters with Cellophane or waxed paper.
To serve put in single layer on baking tray, thaw and heat at 375°F (Gas Mark 5) for 10 minutes. Toss in sugar before serving.
Storage Time 1 month.

FRUIT CRUMBLE

1 lb. apples, plums or rhubarb
6 oz. plain flour
3 oz. brown sugar
4 oz. butter or margarine

Clean and prepare fruit by peeling and/or slicing and arrange in greased pie dish or foil container, sweetening to taste (about 3 oz. sugar to 1 lb. fruit). Prepare topping by rubbing fat into flour and sugar until mixture is like breadcrumbs. Sprinkle topping on fruit and press down.
Pack the fruit crumble uncooked by covering with heavy duty foil, or by putting container into polythene bag.
To serve put container into cold oven and cook at 400°F (Gas Mark 6) for 30 minutes, then at 375°F (Gas Mark 5) for 45 minutes.
Storage Time 6 months (apples may discolour and a little lemon juice will help to prevent this).

FREEZER FUDGE

4 oz. plain chocolate	1 lb. sifted icing sugar
4 oz. butter	2 tablespoons sweetened
1 egg	condensed milk

Melt chocolate and butter in a double saucepan over hot water. Beat egg lightly, mix with sugar and condensed milk, and stir in the chocolate mixture. Turn into greased rectangular tin.

Pack fudge by covering container with lid of heavy duty foil. Freeze for 6 hours, cut in squares and repack in Polythene bags. Store in freezer.

To serve leave at room temperature for 15 minutes.

Storage Time 3 months.

Chapter Thirteen

FOOD FOR ONE

It is commonly felt that a home freezer is only a practical proposition for those who cope with bulk buying, bulk feeding or bulk entertaining. In fact a freezer can be an invaluable aid in saving time, fuel and money for the person living alone, and for providing the nutritious meals which are so often missed by the careerist or by the solitary older person.

A person living alone and out at work during the day needs not only an evening meal but also a selection of meals for weekend or holiday use, and sometimes larger special meals for entertaining. With a freezer, this busy person can shop in bulk and save money, can cook in batches at the weekends, and can prepare special meals in advance for entertaining, rather than resort to a constant diet of quick grills and salads. Even large joints of meat become a possibility when slices can be frozen for quick snacks, and the remains converted into meals for other occasions in the future.

Older people living alone find shopping tiring, and tend to skip cooking. They can take advantage of frozen commercially prepared meals, or friends and family can supply home-cooked dishes for the freezer, complete with instructions for quick preparation, and this can be a tremendous help particularly when weather is bad or health uncertain.

Even large families face the single-meal problem, for many women are on their own during the day, or even for five whole days of the week while families are at school or working, so that while the weekend may bring a rush of demands for large meals, there must be many solitary lunches or suppers. These can be catered for by freezing leftovers in single portions, and by preparing special cooked meals in individual containers.

While most of the recipes in this book can be cooked and split up into smaller portions for freezing for single meals, the following ideas will make solitary catering a little easier.

RAW MATERIALS
Large bags of commercially frozen vegetables and fruit can be split up into individual servings and repackaged; chops and steaks can be separated by Cellophane or greaseproof paper and packed in polythene bags; fish portions can be likewise dealt with. Fresh fruit and vegetables, such as raspberries or beans, can be packed in individual one-meal containers. Even for the large family, it is a good idea to pack such produce in a variety of sizes so that large packs are available for weekend use but individual portions are quickly prepared by the solitary housewife.

LEFTOVERS
Fuller details will be found in LEFTOVERS, but briefly leftover meat and vegetables, soups and sauces can be frozen to use as single meals. In practice, this means that a single person can economise in preparation time and fuel by cooking more than is necessary for immediate use and freezing a portion for later serving. A housewife with a family can process and pack her leftovers for single-portion meals during the week. For ease of reheating and serving, a single person can package a complete meal in a compartmented tray made up with leftover meat and vegetables.

BATCH COOKING
The single person can take advantage of batch cooking in the same way that families can. Small cakes and scones can be made in a family-sized batch, then packaged in twos or fours for individual meals. A special casserole can be made in a large quantity, then packed in three or four portions suitable for entertaining. An economical quantity of soup may be made, then frozen in small blocks to give individual servings; a good spaghetti sauce can be treated in the same way. Puddings are particularly worth making in individual portions; perhaps a double or treble serving can be packed for entertaining, and the remainder divided into small containers for single meals. Such accessories as grated cheese or toasted croutons can be prepared in bulk, packed in small quantities and used to liven up meals without a lot of work after a busy day.

Chapter Fourteen

PICNICS

The freezer can be a valuable aid in providing everyday packed meals and more elaborate picnics. Many recipes in STARTERS: PIES, PASTRY AND PIE FILLINGS, and SANDWICHES AND FILLINGS can be prepared in quantity when the warmer weather arrives to provide instant meals for carrying. Many of these can be packed straight from the freezer and will be thawed and ready to eat at the end of a journey. The only additions need be salad vegetables, since even rolls and baps can be taken straight from the freezer, chunks of cheese, and individual puddings or sugared fruit. For colder weather, soup from the freezer can be quickly thawed in a double boiler and put into a Thermos flask for carrying; in the summer, frozen juices will thaw during the journey and will be refreshingly chilled for drinking.

Sandwiches can be frozen in small packets of individual varieties so that a good selection can be taken for a large picnic. Pies can also be individually made, or larger pies made and frozen in wedges for single servings. Meat and chicken pies, meat balls and fried chicken all freeze well (see Index).

Small batches of cakes and baked biscuits are useful for quick packed meals, or larger cakes may be cut in wedges before freezing. Individual portions of puddings, or small packs of sugared raspberries or strawberries, fruit salad in syrup or sweetened fruit puree will give variety to packed meals.

In addition to suitable recipes which will be found in the Index, the following are particularly good for picnic use.

BEEF GALANTINE

1 lb. chuck steak
4 oz. bacon
4 oz. fine white breadcrumbs
1 teaspoon chopped parsley
1 teaspoon chopped thyme
Salt and pepper
2 eggs

Mince steak and bacon and mix with other ingredients, moistening with egg. Put into a loaf tin and steam for 3 hours. Cool under weights and turn out.
Pack in foil as a whole item; or cut in slices, separating with Cellophane or greaseproof paper and then wrap in foil as a complete item.
To serve thaw in refrigerator overnight, or thaw for 3 hours before packing for journey. Individual slices may be taken out and thawed on absorbent paper, or put into baps or rolls.
Storage Time 1 month.

BACON LOAF

1 lb. cold boiled bacon	1 tablespoon chutney
6 oz. corned beef	1 teaspoon grated lemon rind
4 oz. fresh white breadcrumbs	Salt and pepper
1 small onion	1 egg
1 tablespoon chopped parsley	2 tablespoons milk

Mince bacon and corned beef, and mix with breadcrumbs, finely chopped onion, and all other ingredients. Adjust seasoning according to the saltness of the meat. Pack into greased loaf tin and bake at 350°F (Gas Mark 4) for 1¼ hours. Cool completely and turn out of tin. For sandwiches or baps, the mixture may be cooked in two cocoa tins, and then can conveniently be used in round slices.
Pack in foil.
To serve thaw in refrigerator overnight, or for three hours before packing for journey.
Storage Time 1 month.

SAUSAGE AND BEEF ROLL

8 oz. minced fresh beef	Salt and pepper
8 oz. pork sausage meat	Pinch of nutmeg
4 oz. streaky bacon	2 tablespoons tomato ketchup
3 oz. fresh white breadcrumbs	1 egg
4 tablespoons stock	

Mix beef and sausage meat and add bacon cut in small pieces. Mix with breadcrumbs, stock, seasonings, ketchup and egg. Put

into loaf tin. Cover with greased paper and bake at 375°F (Gas Mark 5) for 1½ hours. Cool and press until firm.
Pack in foil.
To serve thaw in refrigerator overnight, or for 3 hours before packing for journey.
Storage Time 1 month.

SAUSAGE AND ONION PIE

8 oz. short pastry	1 egg
8 oz. pork sausage meat	1 teaspoon mixed herbs
1 onion	

Line foil pie plate with pastry. Mix sausage meat, finely chopped onion, egg and herbs, and put into pastry case. Cover with pastry lid, seal firmly, brush with beaten egg mixed with a pinch of salt, and bake at 425°F (Gas Mark 7) for 30 minutes. Cool completely.
Pack by wrapping in foil.
To serve thaw for 3 hours at room temperature or on journey.
Storage Time 1 month.

BACON PASTIES

12 oz. short pastry	1 large onion
8 oz. minced raw steak	Salt and pepper
6 oz. streaky bacon	½ teaspoon Worcestershire sauce
4 oz. lambs kidney	

Roll out six 7 ins. pastry rounds. Chop all ingredients finely and mix well together. Put a spoonful of mixture on each pastry round and form into pastry shapes, sealing edges well. Place on wetted baking sheet and bake at 425°F (Gas Mark 7) for 15 minutes, and then at 350°F (Gas Mark 4) for 45 minutes. Cool.
Pack in foil tray in polythene bag, or in individual bags.
To serve thaw 2 hours at room temperature or on journey.
Storage Time 1 month.

CORNED BEEF ENVELOPES

4 oz. corned beef	1 tablespoon tomato ketchup
8 oz. short pastry	1 teaspoon chopped parsley

Roll out pastry and cut into twelve squares. Mix corned beef, ketchup and parsley. Put a spoonful of mixture on each square, and fold into triangles, sealing edges well. Brush with egg or milk. Bake at 425° F (Gas Mark 7) for 15 minutes. Cool.
Pack in foil tray in polythene bag, or in shallow box.
To serve thaw at room temperature for 1 hour, or on journey.
Storage Time 1 month.

PICNIC TEA LOAF

1 lb. mixed dried fruit	1 egg
8 oz. sugar	2 tablespoons marmalade
½ pint warm tea	1 lb. self raising flour

Soak fruit with sugar and tea overnight. Stir egg and marmalade into fruit and mix well with flour. Pour into two 1 lb. loaf tins and bake at 325°F (Gas Mark 3) for 1¾ hours. Cool in tins for 15 minutes before turning out. Cool.
Pack in polythene bags or foil.
To serve thaw in wrappings for 3 hours, slice and butter.
Storage Time 4 months.

RAISIN SHORTCAKE

4 tablespoons orange juice	2 oz. caster sugar
4 oz. seedless raisins	4 oz. butter
6 oz. plain flour	

Put orange juice and raisins into a pan and bring slowly to the boil; leave until cold. Sieve flour into a basin and work in the sugar and butter until the mixture looks like fine breadcrumbs. Knead well and divide dough into two pieces. Form into equal-sized rounds. Put one on a greased baking sheet, spread on raisin mixture and top with second round of dough, pressing together firmly and pinching edges together. Prick well. Bake at 350°F (Gas Mark 4) for 45 minutes, mark into sections, and remove from tin when cold.
Pack in foil.
To serve thaw in wrappings for 3 hours.
Storage Time 4 months.

HONEY LOAF

4 oz. butter
4 oz. caster sugar
6 tablespoons honey
1 egg
10½ oz. plain flour
3 teaspoons baking powder
1 teaspoon salt
½ pint milk

Cream butter and sugar until light and fluffy, and mix in honey thoroughly. Beat in egg. Sieve flour, baking powder and salt, and stir into creamed mixture alternately with milk. Put into greased 2 lb. loaf tin and bake at 350°F (Gas Mark 4) for 1¼ hours. Cool.

Pack in polythene bag or foil.

To serve thaw in wrappings for 3 hours, slice and spread with butter.

Storage Time 4 months.

Chapter Fifteen

DINNER PARTIES

Preparing a dinner party can be a nerve-wracking business, for accurate timing can be thrown out by late arrivals or hard drinking guests, and the hostess is often under scrutiny from barely-known business acquaintances, or from critical in-laws or new neighbours. Thankfully, many of the old rules about the number of required courses and the type of food served on formal occasions are being quietly scrapped. In the days when etiquette forbade the discussion of food, the hostess had to adhere slavishly to conventional ideas, and would in some households never have dared to prepare a casserole for guests, start an evening meal with a foundation of soup, or offer a cheese course. Today, a three course meal is acceptable, sometimes with the addition of cheese or fruit, and most dinner party food is only a slightly more elaborate version of the same meals offered to the family.

A wise hostess does not attempt to offer three hot courses, and for ease of serving many now offer a cold first and third course with a hot main course. In the summer, the pattern may be varied and a hot soup precede a cold meal, or a hot pudding finish one. The freezer is invaluable for dinner party preparation, for a complete meal may be assembled at leisure and frozen weeks ahead, or one or two courses may be thus organised, leaving the hostess free to concentrate on the tricky preparation of only one course, or of some special seasonal food.

It is as well to remember when preparing a special dinner that commercially frozen foods are recognisable by many people today, and a better illusion will be created by serving home-made frozen dishes with fresh vegetables in season, or with fresh herbs as a garnish, or a salad; bulk ice-cream too will be better converted into a more elaborate pudding than served straight in the way a home-made ice or sorbet can be offered.

Almost all the recipes in this book can be used for dinner parties. It is always worth tasting them carefully before serving, adjusting seasonings, and perhaps adding a little cream, wine or sherry as appropriate. The vital point to remember about planning a dinner party around frozen food is that while a great deal of last minute preparation and strain is avoided, there must be great care in organising thawing and heating preparations so that the meal is served in perfect condition. To illustrate this, the following menu is detailed not only in recipe preparation but in final assembly. Other menus should be timed in the same way to ensure perfect results.

DINNER PARTY MENU

COQUILLES ST. JACQUES
COQ AU VIN
FRESH VEGETABLES IN SEASON
DUCHESSE POTATOES
STUFFED PEACHES

COQUILLES ST. JACQUES

8 scallops (fresh or frozen)
½ pint dry white wine
1 small onion
Parsley, thyme and bayleaf
4 oz. butter

Juice of 1 lemon
4 oz. small mushrooms
1 tablespoon cornflour
Salt and pepper
2 oz. grated cheese

Clean scallops and put in a pan with wine, chopped onion and herbs. Simmer 5 minutes, no longer as they become tough, and drain scallops, keeping liquid. Melt half the butter, add lemon juice and cook sliced mushrooms until just soft. Drain mushrooms. Add remaining butter to pan and pour in liquid from scallops. Simmer 2 minutes, then thicken with cornflour mixed with a little water. Stir in salt and pepper and grated cheese. Cut scallops in pieces, mix with mushrooms and a little sauce, and divide between 8 scallop shells or individual dishes. Coat with remaining sauce. If liked, pipe edges with creamed potato.
Pack by putting shells on to trays, freezing, then wrapping in foil.

To serve heat frozen scallops at 400°F (Gas Mark 6) for 20 minutes, after sprinkling surface with a few buttered breadcrumbs (these may also be frozen).
Storage Time 1 month.

COQ AU VIN

2 3 lb. chickens	1 tablespoon tomato purée
8 oz. bacon	1 pint red wine
20 small white onions	Parsley, thyme and bayleaf
2 oz. butter	Pinch of nutmeg
2 oz. oil	12 oz. button mushrooms
2 tablespoons brandy	1 garlic clove
Salt and pepper	1 tablespoon cornflour

Chicken joints may be used for this dish; otherwise clean and joint chickens. Cut bacon into small strips, simmer in water for 10 minutes and drain. Peel the onions. Melt butter and oil and lightly fry bacon until brown. Remove from pan and then brown the onions and remove from pan. Fry chicken joints until golden (about 10 minutes), then add bacon and onions. Cover and cook over low heat for 10 minutes. Add brandy and ignite, rotating the pan until the flame dies out. Add salt and pepper, tomato purée, wine, herbs and nutmeg and crushed garlic and simmer on stove or in oven for 1 hour. Remove chicken pieces and put into freezer container. Cook mushrooms in a little butter and add to chicken pieces. Thicken gravy with cornflour, cool, and pour over chicken and mushrooms.
Pack by covering container with lid or foil.
To serve put chicken and sauce in covered dish and heat at 400°F (Gas Mark 6) for 45 minutes.
Storage Time 1 month.

DUCHESSE POTATOES

2 lbs. cooked potatoes	Salt and pepper
4 oz. butter	Pinch of nutmeg
2 eggs	

Put potatoes through a sieve, and beat well with potatoes and eggs to give a piping consistency, seasoning well. A little hot milk may be added if mixture is stiff. Pipe in pyramids on to baking sheets lined with oiled paper.

Pack frozen shapes into bags of suitable quantities.

To serve put on to baking sheets, brush with egg and cook at 400°F (Gas Mark 6) for 20 minutes.

Storage Time 1 month.

STUFFED PEACHES

8 large ripe peaches	Grated rind of 1 lemon
4 oz. ground almonds	1 tablespoon orange juice
3 oz. icing sugar	4 tablespoons sherry
1½ oz. butter	

Peel peaches and cut in halves. Mix ground almonds with icing sugar, soft butter and lemon rind and work in orange juice to give a soft paste. Form paste into eight small balls. Put peaches together round stuffing and put on oven dish. Pour on sherry and sprinkle thickly with icing sugar. Bake at 400°F (Gas Mark 6) for 15 minutes and cool quickly.

Pack in individual containers, or by covering container in which peaches have been cooked in foil.

To serve thaw at room temperature for 2 hours, uncover and heat at 375°F (Gas Mark 5) for 10 minutes. Serve with cream.

Storage Time 1 month.

DINNER PARTY FREEZER PLAN

2 hrs. before dinner	Remove Stuffed Peaches from freezer
1½ hrs. before dinner	Remove Coq au Vin from freezer and put in oven
20 minutes before dinner	Remove Coquilles St. Jacques and Duchesse Potatoes from freezer and put in oven
15 minutes before dinner	Begin cooking fresh vegetables
After first course	Put Stuffed Peaches in oven

Chapter Sixteen

PREPARING FOR CHRISTMAS

While a freezer is always useful for storing prepared food for entertaining, it becomes invaluable for the Christmas cooking marathon. Not only can such raw materials as poultry and vegetables be prepared months in advance when plentiful and cheap, but all the traditional trimmings of Christmas meals can be cooked and stored weeks, and sometimes, months ahead. Additionally, leftovers may be transformed and frozen for later entertaining.

In addition to the items which are dealt with in this chapter, those in STARTERS; SWEET COURSES; CAKES & BISCUITS; and FOOD FOR PARTIES will be useful when planning Christmas menus.

POULTRY

Turkeys and chickens frozen whole will store 8–12 months; geese and ducks 6–8 months. They should be frozen without giblets or stuffing. Roast birds to be served cold are not very successful, as the flesh exudes moisture and becomes flabby on thawing; also there is little time saved in preparation.

STUFFING

The storage life of stuffing is only 1 month, and it is best packed separately. In the normal way, freezing stuffing is unnecessary, since it can be prepared while the poultry is thawing, but at busy times, frozen stuffing can be useful. If it is vital to stuff the bird in advance, a breadcrumb stuffing can be used for a fresh bird, and the stuffed bird frozen for up to 4 weeks; a sausage stuffing should not be used for this.

GIBLETS

Giblets have a storage life of 2 months, and are best packed separately and frozen rather than packed in the bird. Livers may be packed separately for use in paté or omelettes. Giblets can be cooked and the stock frozen for gravy and soup; the chopped giblets can be frozen covered in stock and stored up to 4 weeks for inclusion in soup or pies.

THAWING OF POULTRY

Poultry should be completely thawed before cooking. Thaw in the refrigerator for even slow thawing. A 4–5 lb. chicken will thaw overnight in a refrigerator, and in 4 hours at room temperature. A 9 lb. turkey will take 36 hours, and as much as 3 days should be allowed for a very large bird. Poultry is best thawed in the unopened freezer wrapping, and should be kept no longer than 24 hours in a refrigerator after thawing.

VEGETABLES

Peas, beans, carrots, sweetcorn and sprouts are particularly useful for Christmas. Potatoes mashed with butter and hot milk can be frozen in bags or cartons and reheated in a double boiler. Croquettes can be made from the mixture and fried before freezing (thaw for 2 hours before heating in a moderate oven for 20 minutes).

CHRISTMAS PUDDINGS

These are normally stored in bowls in a dry place, but modern homes are often deficient in storage space. Rich fruit puddings can be made to traditional recipes, in foil bowls, covered with foil lids, and stored up to 1 year in the freezer.

MINCEPIES

Mincemeat is highly spiced, and spices tend to develop "off flavours" in the freezer, so storage life is no longer than 1 month. Pies may be baked and packed in cartons for freezing. If there is more space, pies may be frozen unbaked in their baking tins. Unbaked pies have better flavour and scent, and crisper and flakier pastry than pies baked before freezing.

BASIC POULTRY STUFFING

2 oz. suet
4 oz. fresh white breadcrumbs
2 teaspoons chopped parsley
1 teaspoon chopped thyme
Grated rind of $\frac{1}{2}$ lemon
Salt and pepper
1 medium egg

Grate suet and mix all ingredients together, binding with the egg. The stuffing may be frozen uncooked, or may be cooked as forcemeat balls. If a bird is to be stuffed for freezing, make sure both bird and stuffing are very cold, and that the bird is stuffed in a very cold place. The freezer life of a bird ready-stuffed will be only that of the stuffing (i.e. 1 month).

Pack stuffing into cartons or polythene bags. Deep-fried forcemeat balls may be packed in cartons or bags.

To serve thaw stuffing enough to use in poultry. Ready-cooked forcemeat balls can be put into a roasting tin with the poultry or into a casserole 10 minutes before serving time.

Storage Time 1 month.

SAUSAGE STUFFING

1 lb. sausage meat
2 oz. streaky bacon
Liver from turkey or chicken
1 onion
1 egg
2 oz. fresh white breadcrumbs
Salt and pepper
2 teaspoons fresh mixed herbs
Stock

Put sausage meat in a bowl. Mince bacon, liver and onion. Mix with sausage meat, egg, breadcrumbs, seasonings and herbs, and moisten with a little stock if necessary. Do not stuff bird in advance with sausage stuffing.

Pack in cartons or polythene bags.

To serve thaw in refrigerator for 12 hours before using to stuff bird.

Storage Time 2 weeks.

CHESTNUT STUFFING

1 lb. chestnuts
2 oz. fresh white breadcrumbs
1 oz. melted butter
2 teaspoons fresh mixed herbs
2 eggs
Salt, pepper and dry mustard

Peel chestnuts, then simmer in a little milk until tender. Sieve and mix with breadcrumbs, butter, herbs and eggs. Add salt and pepper and a pinch of dry mustard.

Pack in cartons or polythene bags.

To serve thaw in refrigerator for 12 hours before stuffing bird.

Storage Time 1 month.

BREAD SAUCE

1 small onion	2 oz. fresh white breadcrumbs
4 cloves	½ oz. butter
½ pint milk	Salt and pepper

Peel onion and stick with cloves. Put all ingredients into saucepan and simmer for 1 hour. Remove onion, beat sauce well, and season further to taste. Cool.

Pack in small waxed containers.
To serve thaw in top of double boiler, adding a little cream.
Storage Time 1 month.

CRANBERRY SAUCE

1 lb. cranberries	¾ lbs. sugar
¾ pint water	

Rinse the cranberries. Dissolve sugar in water over gentle heat, add cranberries and cook gently for 15 minutes until cranberries pop. Cool.

Pack in small waxed containers.
To serve thaw at room temperature for 3 hours.
Storage Time 1 year.

APPLE SAUCE (for duck, goose or pork)

1 lb. apples	Squeeze of lemon juice
4 tablespoons water	Sugar to taste

Cook apples in water; the flavour will be better if they are cooked sliced but unpeeled in a casserole in the oven. Sieve apples and sweeten to taste, adding a squeeze of lemon juice.

Pack in small waxed containers.
To serve heat gently in double boiler, adding a knob of butter.
Storage Time 1 year.

BRANDY BUTTER

2 oz. butter	2 tablespoons brandy
2 oz. icing sugar	

Cream butter and sugar and work in brandy.
Pack in small waxed containers, pressing down well.

To serve thaw in refrigerator for 1 hour before serving with pudding or mince pies.
Storage Time 1 year.

CREAMED TURKEY

Cooked turkey   White sauce

Cut cooked turkey into small neat pieces and bind with white sauce made with half turkey stock and half milk, and thickened with cornflour. Cool.
Pack in waxed containers.
To serve reheat in a double boiler to serve with toast or rice, with the addition of a few mushrooms, peas or pieces of green pepper. Also use as a filling for pies or flan cases.
Storage Time 1 month.

TURKEY ROLL

12 oz. cold turkey	Salt and pepper
8 oz. cooked ham	½ teaspoon mixed fresh herbs
1 small onion	1 large egg
Pinch of mace	Breadcrumbs

Mince turkey, ham and onion finely and mix with mace, salt and pepper and herbs. Bind with beaten egg. Put into greased dish or tin, cover and steam for 1 hour. This may be cooked in a loaf tin, a large cocoa tin lined with paper or a stone marmalade jar. While warm, roll in breadcrumbs, then cool completely.
Pack in polythene or heavy duty foil.
To serve thaw at room temperature for 1 hour, and slice to serve with salads or in sandwiches.
Storage Time 1 month.

Chapter Seventeen

FOOD FOR PARTIES

A great deal of time can be saved by preparing and freezing ahead food for cocktail or buffet parties; much of the food for these occasions is small and fiddly to prepare, but, if the work is spread out over a period, the cook-hostess can emerge unscathed. Even when a large party is not planned, it can be useful to keep in the freezer a small selection of "finger foods" which can be quickly served to the casual visitor who comes for a drink, or which can provide a breather for the occasions when a main meal is delayed and there has to be some fast work in the kitchen.

In addition to party food, advance preparations can be made with additional quantities of ice cubes, blocks of ice for punches and cups, and garnishes for drinks or food.

ICE CUBES

Quantities of ice cubes can be frozen in advance; each cube should be wrapped in foil, then a number of cubes packed in a polythene bag for easy storage. A sprig of mint or borage, a twist of orange or lemon, or a cherry can be frozen into each ice cube. For children's parties, fruit squashes or syrups may be frozen into cubes to use as ices or to add sophistication to drinks. For punches and cups, large blocks of ice are preferable, as they do not melt so quickly and dilute the drink; these may be made in ice cube trays without the divisions, or in foil trays, and stored in polythene bags.

GARNISHES

Sprays of herbs become limp when defrosted after thawing. Parsley sprigs may however be frozen for use with sandwiches or pies, but are best put on to plates straight from the freezer. Mint leaves frozen in small foil packages or in polythene bags are useful for drinks. Strawberries with hulls, or cherries on stalks may also be frozen in polythene bags for use with fruit drinks or puddings (they are best frozen unwrapped on metal trays in a single layer, then packaged for easy storage).

Many ideas for party food will be found in SANDWICHES & FILLINGS; PIES, PASTRY AND PIE FILLINGS; and STARTERS. Additional recipes are included in this Chapter.

SANDWICHES AND CANAPÉS

Sandwiches may be prepared up to 4 weeks ahead, but are best frozen in large slices, then cut into small shapes just before serving at a party. Rolled, pinwheel and checkerboard sandwiches can be prepared for party use (see SANDWICHES AND FILLINGS), but should not be sliced until after thawing. Scandinavian open sandwiches may be prepared with white or brown bread, plenty of butter, and suitable toppings. They are best arranged on a tray, baking sheet or foil-wrapped cardboard, then wrapped in polythene or foil, sealed and frozen; they should be stored no longer than 2 weeks. Small canapés can be made and frozen on the same principle. Both open sandwiches and canapés should be thawed at room temperature for 1 hour before serving.

PASTRY

General rules for freezing pastry items will be found in PIES, PASTRY AND PIE FILLINGS. Miniature pies and turnovers may be made and frozen unbaked; they can then be baked before thawing at 400°F (Gas Mark 6) for 15–20 minutes and served hot. Small individual quiches and pizza (see Index) may be baked before freezing and reheated at 350°F (Gas Mark 4) before serving. Sausage rolls, frozen unbaked, can be put straight into the oven and baked at 400°F (Gas Mark 6) for 20 minutes. Vol-au-vent cases may be frozen when baked, but have to be carefully packed to avoid crushing and flaking; it is preferable to make and freeze them unbaked, or to buy them in bulk from a frozen food supplier. If fillings are prepared and frozen separately, these can be heated while the vol-au-vents are being freshly baked.

DIPS AND SPREADS

Dips for dunking crisps or raw vegetables, and spreads for biscuits or stuffed celery, can be frozen if based on cottage or cream cheese. Salad dressing, mayonnaise, hard-cooked egg whites or crisp vegetables should be omitted before freezing, but can be added during thawing time, and packages should be labelled with instructions for finishing. Flavourings such as garlic, onion and bacon may be included in frozen dips and spreads, but they should be carefully packaged to avoid leakage of flavours to other foods in the freezer. Dips and spreads should be thawed for 5 hours at room temperature.

BACON WRAPS

Small bacon-wrapped items can be prepared and frozen, to be grilled or heated in a hot oven after removal from the freezer, until the bacon is crisp. Suitable fillings are prunes stuffed with cream cheese, olives, chicken livers or cocktail sausages, to be wrapped in thin slices of streaky bacon without rinds. These should be secured with cocktail sticks and frozen quickly on a tray, then packaged in polythene bags for storage.

CHEESE CIGARETTES

2½ tablespoons butter
2 tablespoons plain flour
¾ cup creamy milk
¼ teaspoon salt
½ lb. Parmesan cheese
2 egg yolks
¼ teaspoon Cayenne pepper
3 loaves thinly-sliced bread

Melt butter, blend in flour and gradually add milk and salt. Cook on low heat, stirring well, until thick. Remove from heat and stir in cheese, beaten egg yolks and Cayenne pepper. Put in refrigerator in covered bowl, and leave to cool to spreading consistency. Remove crusts from bread slices and flatten each slice with rolling pin. Spread with cheese paste and roll like cigarettes. Enough for 60 to 70 "cigarettes".
Pack after freezing unwrapped on trays, into polythene bags.
To serve thaw at room temperature for 1 hour and fry in deep fat (temperature 475°F) until brown. Drain on absorbent paper and serve hot.
Storage Time 1 month.

CRAB AND CHEESE ROLLS

4 oz. butter
8 oz. cream cheese
1 lb. fresh or canned crabmeat
2 loaves thinly-sliced bread

In a double boiler, melt butter and blend in cream cheese until just warm. Cool and add crabmeat. Remove crusts from bread slices, roll each slice with a rolling pin, and spread with crab and cheese mixture. Roll up like cigarettes and cut each roll in half. Enough for 70 to 80 rolls.

Pack after freezing unwrapped on trays, into polythene bags.

To serve put rolls on baking tray, brush with melted butter, thaw for 30 minutes at room temperature and bake at 400°F (Gas Mark 6) for 10 minutes.

Storage Time 2 weeks.

CHEESE TOASTS

8 oz. Cheddar cheese	1 teaspoon cream
8 rashers grilled lean bacon	1 teaspoon dry mustard
1 medium onion	2 loaves thickly-sliced bread

Mince together cheese, bacon and onion and mix with cream and mustard. Cut bread into rounds or slices without crusts, and toast on one side only. Spread mixture on other side. Enough for 60 pieces of toast.

Pack after freezing unwrapped on trays into foil or polythene.

To serve thaw at room temperature for 1 hour, and grill under medium heat for 4 minutes.

Storage Time 2 weeks.

CREAM CHESE BALLS

Cream cheese	Salt and pepper
Walnuts or onions	Chopped nuts or crushed crisps

Mash cream cheese with finely chopped walnuts or onions and season well with salt and pepper. Roll balls about ¾ ins. diameter, and roll in chopped nuts or crushed crisps. Arrange on foil trays or foil-wrapped cardboard.

Pack with foil overwrap, or in polythene bag.

To serve thaw at room temperature for 1 hour and put on to cocktail sticks.

Storage Time 1 month.

COCKTAIL GRAPES

White grapes	Grated onion
Cream cheese	Salt
Roquefort cheese	Worcestershire sauce

Split grapes and remove seeds. Blend equal parts cream cheese and Roquefort cheese and season highly with grated onion, salt and Worcester sauce. Stuff grapes with filling and arrange on foil trays or foil-wrapped cardboard.
Pack with foil overwrap, or in polythene bag.
To serve thaw at room temperature for 1 hour.
Storage Time 2 weeks.

BLUE CHEESE SPREAD

8 oz. Danish Blue cheese   2 tablespoons chopped parsley
4 oz. soft butter   1 small onion
2 tablespoons port

Blend together cheese, butter, port and parsley with a fork (put bowl in pan of hot water to make this easier). Add minced or grated onion.
Pack in small waxed or rigid plastic containers.
To serve thaw in refrigerator for 3 hours and spread on small salted biscuits.
Storage Time 2 weeks.

CREAM CHEESE AND LIVER SPREAD

1 lb. cream cheese   Worcestershire sauce
8 oz. liver sausage   Salt and pepper

Blend together cream cheese and liver sausage until smooth, and season well with sauce, salt and pepper.
Pack into small waxed or rigid plastic containers.
To serve thaw in refrigerator for 3 hours, and serve on toast or biscuits, as a canapé spread, or as a sandwich filling.
Storage Time 2 weeks.

CRAB AND CHEESE DIP

2 oz. Danish Blue cheese   1 clove garlic
2 oz. cream cheese   1 teaspoon lemon juice
½ teaspoon Worcestershire sauce   6 oz. fresh or tinned crabmeat

Blend together cheeses and gradually work in sauce, crushed garlic, lemon juice and crabmeat.
Pack in waxed or rigid plastic container.

To serve thaw in refrigerator for 3 hours and serve in a bowl surrounded by crisps.
Storage Time 2 weeks.

HAM AND HORSERADISH DIP

8 oz. cooked ham	1 tablespoon grated horseradish
1 tablespoon chopped parsley	4 oz. cream cheese

Put ham through mincer and blend in parsley, horseradish and cream cheese.
Pack in waxed or rigid plastic container.
To serve thaw in refrigerator for 3 hours and serve in a bowl surrounded by crisps or small salted biscuits.
Storage Time 2 weeks.

ORANGE CHEESE DIP

4 oz. cream cheese	¼ teaspoon salt
1 tablespoon grated orange rind	Pinch of paprika

Blend together cheese, orange rind, salt and paprika.
Pack in waxed or rigid plastic container.
To serve thaw in refrigerator for 3 hours, and serve in a bowl surrounded by crisps.
Storage Time 2 weeks.

Chapter Eighteen

FOOD FOR BABIES AND SMALL CHILDREN

It is extremely useful to be able to keep small quantities of special food for babies, small children, and even old people in the freezer. There is no nutritional loss from freezing foods. In fact, one is able to preserve the vitamin content more effectively by freezing fruit and vegetables when absolutely fresh. Fruit and vegetable purée or mixtures of vegetables and meat, can be frozen in ice cube trays, and the cubes individually wrapped for storage. Each cube will provide a portion for a baby. This means that reasonable quantities of such foods as carrots, mixed vegetables or spinach can be prepared at one time, instead of making individual servings for each meal which is wasteful of time, fuel and ingredients.

It is of course particularly important that a very high standard of hygiene is observed in preparing these foods, and that the food is cooled rapidly in a germ-free atmosphere, freezing takes place rapidly, and reheating is quick and thorough.

The freezer can also be used to keep portions of commercially canned baby foods – if a tin is opened and half-used, the remains can be quickly frozen by the ice cube method for use at a later date.

For toddlers and old people, slightly more solid food can be prepared such as creamed chicken or fish, creamed minced beef and individual portions of mashed potatoes or vegetable purée, and this can be a good way of turning family leftovers into useful secondary meals. Individual puddings are also useful for this type of meal, and it is a great help for housewives who have to prepare large evening meals to have these rather fiddly luncheon meals taken care of in the freezer for rapid heating and service.

Useful recipes for fruit syrups suitable for children will be found in PRESERVES FROM THE FREEZER; many of the recipes in SWEET COURSES can also be adapted to individual portions, and the chapter on STARTERS has recipes for soups which can be packed in individual portions.

CHICKEN IN CREAM SAUCE

1 boiling chicken	Parsley and bayleaf
1 small onion	Salt, pepper, nutmeg
1 clove	1 tablespoon cornflour
1½ pints milk	

Put chicken in a deep casserole and pour in milk, together with the onion stuck with clove, herbs and seasonings. Cover and cook at 300°F (Gas Mark 2) for 3 hours. Remove chicken and cut in thin slices, or mince flesh. Strain milk and thicken with cornflour, adjusting seasoning. Mix chicken and sauce, and cool.

Pack in small portions in waxed or rigid plastic containers or in foil containers. Thicker slices in sauce may be packed in larger portions for family use.

To serve reheat in double boiler.

Storage Time 1 month.

FISH PIE

1½ lbs. cod fillet	8 oz. mashed potato
½ pint white sauce made with cornflour	

Optional – 2 oz. shelled prawns or peas

Poach cod until just cooked, and flake flesh. Put in dish with prawns or peas if used and cover with sauce. Cover with potato mashed with a little milk. Brush surface with a little beaten egg and bake at 400°F (Gas Mark 6) for 35 minutes. Cool.

Pack by putting dish into polythene bag or foil. This may be made in individual dishes if preferred.

To serve heat at 350°F (Gas Mark 4) for 20 minutes.

Storage Time 1 month.

CREAMED SPINACH

2 lbs. spinach	4 tablespoons thin cream or top milk
Salt and pepper	
1 oz. butter	1 teaspoon lemon juice
½ oz. cornflour	

Wash spinach thoroughly, and cook with a little salt until tender. Drain thoroughly and chop finely. Melt butter and stir in milk blended with cornflour. Season well and simmer for 2

minutes. Add chopped spinach and lemon juice and heat through. Cool.
Pack in small portions in ice cube trays or individual containers
To serve heat in double boiler.
Storage Time 1 month

CREAMED POTATOES

| Cooked potatoes | Salt and pepper |

Whip up potatoes while still hot and season lightly. Cool.
Pack in individual cartons.
To serve heat a little milk in double boiler, stir in potato and beat while it thaws.
Storage Time 1 month.

FRUIT WHIP

| 1 lb. rhubarb or gooseberries or blackcurrants | Sugar
½ pint evaporated milk |

Prepare fruit and stew in very little water with sugar to taste until tender. Put through a sieve and fold into whipped evaporated milk. This may be made with whipped cream, but the tinned milk is less rich for small children or old people.
Pack in individual dishes and cover with foil.
To serve thaw at room temperature for 2 hours.
Storage Time 2 months.

ORANGE CASTLES

4 oz. butter	4 oz. self raising flour
4 oz. sugar	Grated rind of 1 orange
2 eggs	3 oz. sultanas

Cream butter and sugar until light and fluffy. Beat in eggs one at a time with a little flour. Stir in orange rind and a little milk to moisten together with rest of flour. Put into 8 individual buttered castle pudding moulds, cover tops with foil and steam for 45 minutes. Cool, and remove from moulds.
Pack in foil, or leave in moulds and cover with foil.
To serve reheat in foil in oven or steamer for 20 minutes and serve with custard or a little jam.
Storage Time 2 months.

Chapter Nineteen

PRESERVES FROM THE FREEZER

Time is often limited for preparing jams, pickles and syrups, but when the raw materials are available in the garden, or cheap in the shops, it is worth varying the store cupboard with a small selection of preserves to augment meals. It is particularly useful in a glut season of raspberries or blackcurrants, for instance, to turn some of them into syrup, which saves on freezer space, and at the same time provides a time-saving step towards the making of drinks, puddings or sauces.

Modern store cupboards have little room for these extra supplies, but they take up little space in the freezer; in addition, really fresh fruit flavours are preserved, and there is no danger of preserves drying out, crystallising or tasting "winey". Frequently cooking time is saved in making preserves for the freezer, and the messy chore of bottling, covering and sterilising can be avoided.

FRUIT SYRUPS

Any standard syrup recipe may be used, but specific details are included in this chapter. Syrups may be frozen in small waxed or rigid plastic containers. Syrup can also be frozen in ice-cube trays, then each cube wrapped in foil and packed in quantities in bags for easy storage; one cube will then give an individual serving to use with puddings or ice cream, or to dissolve in water as a drink.

The rules for making fruit syrups are similar to those for jelly-making. Gentle heat should be used to extract the juice, with a little water added to the fruit. A jelly bag should be used to let the juice drip through, and this is then measured and the appropriate quantity of sugar added and dissolved by stirring, before freezing in the preferred containers. When time is valuable, the fruit may be put through a fine sieve, sweetened and frozen as purée. While this is not suitable for drinks, it can be used as a quick sauce for puddings or ice cream, and as a basis for mousses, etc. This is particularly successful with raspberries and blackcurrants. A little colouring may be added to deepen the colour of pale syrups.

FRUIT JUICES

Citrus fruit, apples and tomatoes make excellent juice to store in the freezer. They may be frozen in rigid containers or in ice-cube trays.

JAMS

Uncooked jams for the freezer are quickly and easily prepared, with no tedious boiling and testing, and will store for up to six months. They should be packed in small containers to be used at one serving. If these jams are stiff, or if "weeping" has occurred at the time of serving, they can be lightly stirred to soften and blend. Colour and flavour of these jams will be particularly good.

PICKLES, CHUTNEYS AND RELISHES

One or two special "sharp" accompaniments to savoury dishes may be successfully stored in the freezer, and are useful for the completion of special meals, and for holiday times. Since these usually contain spices which quickly develop "off-flavours" they are best stored for no longer than one month.

SOFT FRUIT SYRUP

Raspberries, blackcurrants, redcurrants, strawberries, blackberries or elderberries
Sugar

Fruit may be used singly, or combined, e.g. raspberries and redcurrants. Use fresh clean ripe fruit and avoid washing if possible, discarding mouldy or damaged fruit. Add ¼ pint water for each lb. of raspberries, strawberries or blackberries; ½ pint for each lb. of currants or elderberries. Cook very gently (This can be done in the oven in a covered jar) for about 1 hour, crushing fruit at intervals. Turn into jelly bag or clean cloth, and leave to drip overnight. Measure cold juice and add ¾ lb. sugar to each pint of juice. Stir well until dissolved.

Pack into small waxed or rigid plastic containers, leaving ½ ins.

headspace. Or pour into ice cube trays and wrap each cube in foil after freezing.
To serve thaw at room temperature for 1 hour, and use for sauces, mousses or drinks.
Storage Time 1 year.

ROSE HIP SYRUP
2½ lbs. ripe red rose hips   1¼ lbs. sugar
3 pints water

Wash rose hips well and remove calyces. Put through mincer and pour on boiling water. Bring to the boil, then remove from heat and leave for 15 minutes. Strain through jelly bag or cloth overnight. Measure juice and reduce to 1½ pints by boiling. Add sugar, stir well to dissolve and boil hard for 5 minutes. Leave until cold.
Pack by pouring into ice cube trays and wrapping cubes in foil when frozen.
To serve thaw at room temperature for 1 hour.
Storage Time 1 year.

APPLE JUICE
2 lbs. apples   ½ pint water

This juice may also be made by simmering leftover peelings in water. Simmer cut-up apples in water very gently, and allow to drip in jelly bag or cloth overnight. Do not sweeten before freezing.
Pack in waxed or rigid plastic containers, leaving ½ ins. headspace.
To serve thaw at room temperature for 1 hour and sweeten to taste.
Storage Time 1 year.

CITRUS FRUIT JUICE
Oranges, grapefruit, lemons or limes
Use good quality fruit which is heavy in the hand for its size. Chill unpeeled fruit in ice water or in the refrigerator until ready to extract fruit juice. Squeeze juice and strain.
Pack in waxed or rigid plastic containers, leaving ½ ins. head-

space. Lemon and lime juice may be frozen in ice cube trays, each cube being wrapped in foil when frozen.
To serve thaw at room temperature for 1 hour. Sweeten to taste. Small cubes of lemon or lime juice are very useful for individual drinks.
Storage Time 1 year.

TOMATO JUICE
Ripe tomatoes

Core and quarter ripe tomatoes and simmer in covered pan for 10 minutes. Drip through muslin, cool before freezing.
Pack in waxed or rigid plastic container
To serve thaw at room temperature for 1 hour. Season to taste with salt, pepper and lemon juice.
Storage Time 1 year.

STRAWBERRY JAM
1½ lbs. strawberries   4 fl. oz. liquid pectin
2 lbs. caster sugar

Mash or sieve strawberries and stir with sugar in a bowl. Leave for 20 minutes, stirring occasionally, then add pectin and stir for 3 minutes.
Pack in small waxed or rigid plastic containers, cover tightly and seal. Leave at room temperature for 24–28 hours until jelled before freezing.
To serve thaw at room temperature for 1 hour.
Storage Time 6 months.

RASPBERRY JAM
1½ lbs. raspberries   4 fl. oz. liquid pectin
3 lbs. caster sugar

Mash or sieve raspberries and stir with sugar. Leave for 20 minutes, stirring occasionally, then add pectin and stir for 3 minutes.
Pack in small waxed or rigid plastic containers, cover tightly and seal. Leave at room temperature for 24–28 hours until jelled before freezing.
To serve thaw at room temperature for 1 hour.
Storage Time 1 year.

BLACKBERRY JAM

1½ lbs. blackberries
2¾ lbs. caster sugar
4 fl. oz. liquid pectin

This is best made with large cultivated blackberries, as the small hard wild ones are difficult to mash without liquid and are rather "pippy". Mash berries and stir into sugar. Leave for 20 minutes, stirring occasionally, then add pectin and stir for 3 minutes.
Pack in small waxed or rigid plastic containers, cover tightly and seal. Leave at room temperature for 24–48 hours until jelled before freezing.
To serve thaw at room temperature for 1 hour.
Storage Time 1 year.

APRICOT OR PEACH JAM

1½ lbs. ripe fresh apricots or peaches
2 lbs. caster sugar
4 fl. oz. liquid pectin
1 teaspoon powdered citric acid

Skin apricots or peaches and remove stones. Mash fruit and stir in sugar and acid. Leave for 20 minutes, stirring occasionally, then add pectin and stir for 3 minutes.
Pack in small waxed or rigid plastic containers, cover tightly and seal. Leave at room temperature for 24–48 hours until jelled before freezing.
To serve thaw at room temperature for 1 hour.
Storage Time 1 year.

RASPBERRY SAUCE

Raspberries
Sugar

Put raspberries in pan with very little water and heat very slowly until juice runs. Put through a sieve and sweeten to taste.
Pack into small waxed or rigid plastc containers.
To serve thaw in container in refrigerator for 2 hours. Serve with puddings or ice cream.
Storage Time 1 year.

PEACH SAUCE

1 lb. ripe peaches
4 oz. caster sugar
1 tablespoon lemon juice

Peel and stone peaches and crush fruit with a silver fork. Put through sieve and mix with lemon juice and sugar.
Pack into small waxed or rigid plastic containers.
To serve thaw in container in refrigerator for 2 hours. Serve with puddings or ice cream.
Storage Time 1 year.

SPICED APPLES

8 oz. sugar
¾ pint water
4 ins. cinnamon stick
6 firm eating apples
Pink colouring

In a saucepan put sugar, water and cinnamon stick, and heat until sugar dissolves. Peel and core apples and cut in ¼ ins. rings. Simmer a few apple rings at a time until just tender, adding a little colouring to syrup for a better appearance. Take slices from syrup, drain and cool. Arrange slices on baking sheet and freeze.
Pack frozen apple slices in heavy duty foil or small containers. Pack syrup (after removing cinnamon) in small containers.
To serve thaw at room temperature for 1 hour, and serve rings with pork, ham or goose. Syrup may be used for making apple sauce, or for adding to pies.
Storage Time 1 month.

CRANBERRY ORANGE RELISH

1 lb. fresh cranberries
2 large oranges
1 lb. sugar

Mince together cranberries and orange flesh, and stir in sugar until well mixed.
Pack in small containers for one-meal servings.
To serve thaw at room temperature for 2 hours. Very good with pork, ham or poultry.
Storage Time 1 year.

Chapter Twenty

COMPLETE MEALS FROM THE FREEZER

Normally, either a main course or a sweet course can be taken from the freezer to save cooking time, and supplemented by fresh food in season for other courses. Or a freshly-cooked main course may be accompanied by frozen vegetables and pudding, or preceded by a soup made from frozen stock.

Sometimes, however, a complete meal may be needed from the freezer, during holiday times, or for unexpected guests when fresh materials are not readily available.

The best plan for these occasions is to see that there is always a good variety of each type of cooked food in the freezer together with appropriate vegetables and fruit in the raw state, and when freezer contents are carefully recorded it need be only a matter of seconds to assemble a complete meal.

It can happen however that a series of ready-meals is wanted for a holiday period or long weekend with guests, or when the usual cook is likely to be away on holiday or perhaps in hospital. It is then a good plan to pack these complete meals together and store them separately so that they can be found speedily. The organised cook can simply put all the components of a meal into one large polythene bag, still in their individual wrappings, and clearly label the parcel for when it is intended. It is particularly important if these dishes are to be finished by someone else that full thawing, heating and seasoning instructions should be added to each dish or a working plan added to each complete parcel.

It might be thought that the answer lies in the preparation of a number of "dinners on a tray" for such occasions, but these are not normally suitable for entertaining guests, and are difficult to assemble. The meals assembled on this basis for commercial sale are carefully and scientifically planned so that the same oven heat will warm meat and vegetables or bake pastry and roast potatoes. Home planners can use these trays for single meals (see LEFTOVERS) but will not find them easy to organise on a large scale, and it is therefore more practical to combine whole cooked dishes.

The following suggestions for menus to suit a variety of occasions have been compiled from the detailed recipes given in this book, to be supplemented by fresh or frozen vegetables as required.

CHICKEN TETRAZZINI STRAWBERRY MOUSSE	TOMATO SOUP STEAK AND KIDNEY PIE BUTTERSCOTCH ICE
MEAT BALLS, TOMATO SAUCE AND RICE FRUIT FRITTERS	COD'S ROE PÂTÉ BEEF IN WINE RASPBERRY AND APPLE PIE
MEAT LOAF BAKED STUFFED POTATOES BAKED APPLE DUMPLINGS	PHEASANT IN CIDER ICEBOX CAKE
KIDNEY SOUP QUICHE LORRAINE APPLE ICE CREAM	PORK WITH ORANGE SAUCE BLACKCURRANT FLAN
CHICKEN IN CURRY SAUCE FRUIT CREAM	MACARONI CHEESE PEARS IN RED WINE
SCOTCH BROTH COTTAGE PIE FRUIT MOUSSE	BROWN VEGETABLE SOUP FISH TURNOVERS RUM AND DATE ICE CREAM

ONION SOUP
OVEN-FRIED CHICKEN
PEACHES IN WHITE
  WINE

SHRIMP BISQUE
CHICKEN PIE
ORANGE SORBET

JUGGED HARE
LEMON ICE PIE

OXTAIL SOUP
FISH CAKES
BAKED CHEESECAKE

VEAL IN TOMATO
  SAUCE
SWEDISH APPLECAKE

FISH PIE
CHOCOLATE MOUSSE

SAVOURY PANCAKES
STRAWBERRY WATER
  ICE

GNOCCHI
KIDNEYS IN WINE
LEMON PUDDINGS

SPAGHETTI WITH
  SPAGHETTI SAUCE
PRALINE ICE CREAM

CHICKEN LIVER PÂTÉ
JELLIED BEEF
COFFEE PUDDING

KEDGEREE
DANISH CHERRY TART

LIVER CASSEROLE
NESSELRODE MOUSSE

PHEASANT PÂTÉ
SPANISH RICE
CARAMEL ICE CREAM

VEAL WITH CHEESE
CUMBERLAND TART

Chapter Twenty-One

HOW TO USE FROZEN FOOD

No home freezer is really complete without a selection of commercially frozen foods. Vegetables and fruit in particular, bought in bulk, save shopping time and money; fish is also another valuable freezer item when shops are far away, and many families enjoy such items as fish fingers, prepared chips, beefburgers or pies, and fruit juices. Also useful are eclairs and sponges filled with cream, and a variety of mousses, frozen puddings and ice cream. These can supplement home-prepared frozen dishes, and can also be turned into very good meals in their own right. Since such items as frozen fish may be used to prepare some of the dishes in this book, it may be useful to know the weight of commercially frozen items comparable to market-bought fresh raw materials:

FISH	Weight of frozen fish	Comparable weight of market-bought fish
Cod	13 oz.	1 lb. 12 oz.
Haddock	13 oz.	1 lb. 12 oz.
Plaice	13 oz.	1 lb. 10 oz.
Kippers	7½ oz.	1 lb. 7 oz.

FRUIT JUICE	Weight of frozen juice	Equivalent in fresh fruit
Orange	6 fl. oz.	Juice of 12 oranges
Grapefruit	6¼ fl. oz.	Juice of 6–8 grapefruit

VEGETABLES	Weight of frozen vegetables	Comparable weight of market-bought vegetables
Peas	10 oz.	1 lb. 12 oz.
Broad Beans	10 oz.	2½ lbs.
Brussels Sprouts	10 oz.	1 lb.
Sliced Green Beans	9 oz.	1 lb.
Spinach	12 oz.	1½ lbs.

FROZEN VEGETABLES

Instructions for cooking vegetables by the normal water-in-a-saucepan method are on commercially frozen packets. However, there are more delicious ways of cooking them for full flavour. When potatoes are being cooked, the frozen vegetables can be steamed over them for 5 minutes longer than the normal cooking time. The still-frozen vegetables may be added to casseroles or stews 20 minutes before the end of cooking time. While a joint is roasting, vegetables can be cooked in a foil parcel with a knob of butter, pepper and salt; peas will take 30 minutes and other vegetables 40 minutes. Vegetables may also be cooked in butter in a saucepan, using 1 tablespoon butter and 2 tablespoons water for 10 oz. frozen vegetables with salt, and using a tight-fitting lid; they should be simmered for 20–30 minutes.

VEGETABLE VARIETY

Peas	Cook peas and serve with chopped skinned tomatoes and spring onions fried lightly in butter. *or* cook chopped celery in a little water until tender, then add peas and cook for 5 minutes
Sliced Green Beans	Add crispy fried bacon and chopped, cooked onions to the cooked beans.
Cut Green Beans	Slice 2 oz. mushrooms and toss in butter, then add to cooked beans.

Brussels Sprouts	Make white sauce, add a little chopped cooked celery or chopped onions and pour over cooked sprouts.
Broad Beans	Add cooked beans to a cheese sauce, sprinkle with grated cheese and brown under grill.
Spinach	Add a little horseradish sauce to the cooked spinach.
Broccoli	Fry a little grated onion in butter, add a few drops of lemon juice and pour over cooked broccoli, topping with a few toasted shredded almonds.
Peas and Carrots	Fry chopped onion in butter and add to cooked vegetables.
Sweet Corn	Mix cooked sweet corn with a little double cream.

Three of the most useful commercially prepared foods which can add variety to meals are chicken pieces, fish and canned orange juice and the following recipes may be prepared for immediate use and in some cases frozen again in the form of cooked meals. They have been devised in co-operation with the Birds Eye Kitchen.

CHICKEN IN CREAM MUSHROOM SAUCE

1 oz. butter	2 tablespoons cider
2 frozen quarter chickens	¼ pint double cream
4 oz. mushrooms	Salt and pepper

Thaw chicken, pat dry and fry gently in butter. Put on dish to keep hot. Toss sliced mushrooms in butter, add cider and cream, and stir gently over low heat. Season to taste and pour over chicken.

CHICKEN POSITANO

2 oz. butter
4 frozen quarter chickens
1 medium onion
1 garlic clove
4 oz. quartered mushrooms

16 oz. can tomatoes
¼ pint red wine
2 tablespoons sherry
2 bayleaves
2 tablespoons cornflour

Thaw chicken, pat dry and fry in butter until lightly browned. Remove from pan to keep warm. Fry chopped onion and crushed garlic until soft. Add mushrooms and replace chicken in pan. Add tomatoes, wine, sherry and bayleaves and simmer for 45 minutes. Thicken gray with cornflour mixed with a little water and simmer for 3 minutes. Serve with rice. This dish may be frozen, and reheated in a double boiler or in a casserole in a moderate oven.
Storage Time 1 month.

KIPPER PÂTÉ

6 oz. frozen buttered kipper fillets
1½ oz. softened butter
Pepper and nutmeg

Cook fillets as directed on packet, and turn contents including juices into a bowl. Remove fish skin, and mince or pound kipper and juices finely. Beat in soft butter and season to taste with pepper and nutmeg. Serve with toast.

MUSHROOM STUFFED PLAICE

1 oz. butter
3 tablespoons white breadcrumbs
1 small onion
Salt and pepper
2 oz. mushrooms
2 whole frozen plaice

Melt the butter and lightly fry chopped onion and mushrooms, mixing with breadcrumbs, salt and pepper. Make a cut down the centre of the skinned side of each fish and fill with stuffing. Bake in a lightly greased oven dish at 400°F (Gas Mark 6) for 30 minutes. This is good served with mushroom sauce, which can be quickly made from a tin of condensed mushroom soup.

COD STEAKS WITH CURRY SAUCE

1 oz. butter	2 level teaspoons curry paste
½ pint water or stock	2 teaspoons lemon juice
1 small onion	1 oz. flour
1 skinned and sliced tomato	Pinch of sugar
1 small apple	14 oz. frozen cod steaks
Salt	

Melt the butter, and add finely chopped onion, tomato and apple, and fry gently until browned lightly. Stir in curry paste, lemon juice, flour and sugar and cook gently for 3 minutes, then stir in water or stock and bring to the boil, stirring well. Add the cod steaks, cover and simmer for 20 minutes. Serve with rice.

RUSSIAN FISH PIE

7½ oz. frozen puff pastry	½ teaspoon grated lemon rind
7½ oz. frozen cod fillet	2 teaspoons chopped parsley
¼ pint white sauce	Salt and pepper

Thaw pastry and roll into 10 ins. square. Mix together cooked flaked fish, sauce, lemon rind, parsley and seasoning, and put in centre of the pastry square. Damp the edges and bring the four corners to the centre, forming an envelope. Seal the edges, brush with a little egg or milk and bake in a hot oven (475°F or Gas Mark 8) for 20 minutes. This pie can be frozen in foil for future use and reheated to serve.

ORANGE HONEYCOMB

1 packet orange jelly	2 eggs
¾ pint water	
1 can frozen orange juice, undiluted	

Melt orange jelly with water and leave to cool. Add orange juice and egg yolks. Whisk egg whites lightly and fold gently into orange mixture. Pour into serving dish or individual glasses, and leave in refrigerator to set.

ORANGE SORBET

2 oz. sugar
½ pint water
¼ oz. gelatine
1 can frozen orange juice, undiluted
1 egg white

Dissolve sugar in hot water and leave to cool. Soak gelatine in a little water, stand in pan of hot water until gelatine is syrupy and add to sugar syrup with undiluted orange juice. Stir, pour into ice tray and freeze for 1 hour. Remove from freezer, add beaten egg white, and continue freezing for 2 hours.

ORANGE CRUMB PIE

8 oz. short pastry
2 oz. butter
2 oz. caster sugar
1 egg
½ can frozen orange juice
4 oz. madeira cake crumbs

Line pie plate with pastry. Cream butter and sugar until fluffy, and work in egg yolk. Add thawed orange juice, then cake crumbs, and stiffly whipped egg white. Pour into pastry case and bake at 375°F (Gas Mark 5) for 40 minutes. This pie is also excellent if frozen uncooked (cover pie plate with foil), then put into oven while still frozen and baked at 375°F (Gas Mark 5) for 1 hour.

Chapter Twenty-Two

LEFTOVERS

Prepared dishes which have been frozen and thawed must not be frozen again under any circumstances. However, if frozen raw materials have been made up into dishes, leftovers may be frozen for future use. For example a joint of frozen beef may be cooked for immediate use, and the remains frozen in the form of slices in gravy, meat loaf, rissoles, cottage pie or sandwich spread. Frozen poultry may be cooked and frozen again in the form of pies, spreads, casseroles or stock for soups.

Cooked meat, vegetables and sauces may be frozen as complete individual meals in compartmented foil trays. The meat, sliced and in gravy can go into one compartment, mixed vegetables topped with parsley butter in another, cooked rice in a third compartment. Such a meal can be reheated under a foil cover at 375° F (Gas Mark 5) for 30 to 45 minutes.

Meat, poultry and vegetables can be transformed into delicious dishes for the freezer, but it is important that the leftovers should be converted and frozen as quickly as possible and not left in a larder or refrigerator for some time before processing. Smaller quantities of leftovers may be frozen individually to be used again after thawing in freshly-made dishes.

VEGETABLES
Cooked vegetables which are not incorporated into casseroles for refreezing, can be put into pies or flans, or made into purée with a little stock and frozen as a basis for soups, stews or sauces.

SOUP AND SAUCES
Small quantities of soup may be frozen for later use as individual servings, or for casseroles or pies, or as sauce for meat or fish. Leftover sauce can be frozen in individual portions in ice cube trays to use with meat, fish, vegetables or pasta. Leftover Italian tomato paste may also be frozen in cubes.

MEAT AND POULTRY

Sliced cooked meat and poultry may be frozen, if packed compactly, but is most satisfactory if covered with sauce or gravy. Stuffing from leftover poultry should be frozen separately. Minced or cubed cold meat or poultry can be frozen to use with sauce for pasta, or to incorporate in freshly-made shepherd's pie, or patties made with pastry. Thawing and preparation time will be saved if the leftover meat is prepared immediately after a meal in the form of meat loaf, rissoles or a casserole. Ham and poultry can be turned into paste for toast or sandwiches.

FISH

Leftover fish is rarely worth converting into other dishes, except fishcakes or fish pie, since fish spoils with overcooking which is bound to occur when a frozen dish is reheated. Small quantities of fish may be mashed with butter, a little parsley, or some anchovy essence to make a spread to freeze for toast or sandwiches.

BACON AND HAM

Cooked bacon can be crumbled and frozen, to use later for topping potatoes, fish or cheese dishes, or for mixing into sandwich spreads, or for adding flavour to casseroles. Cooked ham is best chopped or minced and frozen for use in stuffings, casseroles and spreads.

CHEESE

Small pieces of cheese are best grated and frozen to use for toppings; or it may be mixed with breadcrumbs before freezing, for stuffings or toppings. Leftover cheese may also be put into white sauce and frozen, for use with many made up dishes, and to blend vegetables or poultry in pies or flans.

EGG WHITES AND YOLKS

Quantities of leftover egg whites and yolks may be frozen. The whites need no pre-freezing treatment but should be packed into cartons immediately; they are excellent for meringues if used at room temperature. Yolks should be mixed very lightly with a fork, allowing $\frac{1}{2}$ teaspoon salt or $\frac{1}{2}$ tablespoon sugar to 6 yolks, according to possible end use, then packed and carefully labelled. They can be used when thawed exactly as fresh eggs.

BREAD
Ordinary or buttered breadcrumbs are usefully frozen in bags to use for topping dishes; they may be mixed with grated cheese. Bread slices can be frozen and grated into crumbs while still hard. Stale bread cut in small cubes may be toasted, with or without melted butter, and frozen to use as croutons for soup.

CREAM
Surplus whipped cream can be piped or dropped in spoonsful on to baking sheets, fast frozen, then packed into polythene bags. After removing from wrappings, cream will thaw in 30 minutes at room temperature.

COFFEE AND TEA
Surplus strong tea and coffee can be frozen in ice cubes, then each cube wrapped in foil and put in bags for storage. They are useful for iced tea and coffee, which they will not dilute while chilling, and tea cubes are useful for summer fruit cups.

FRUIT PEELS, JUICES AND SYRUPS
Grated lemon and orange peel can be frozen in small waxed or plastic containers and is useful for flavouring cakes, puddings and preserves. Lemon and orange juice can be frozen in ice cubes, then packed in foil for storage. Surplus syrup from canned fruits will freeze in cubes and is good for fruit drinks and as the basis of fruit sauces to use with puddings and ice cream.

STUFFED CABBAGE ROLLS
1 lb. minced cold meat
1 oz. butter
1 small onion
2 tablespoons cooked rice
1 teaspoon chopped parsley
Salt and pepper
Stock
12 medium-sized cabbage leaves

Cook meat in butter together with the onion until the meat begins to colour. Mix with rice, parsley, salt and pepper, and enough stock to moisten, and cook for 5 minutes. Blanch cab-

bage leaves in boiling water for 2 minutes and drain well. Put a spoonful of filling on each leaf, and form into a parcel, and put parcels close together in a covered oven dish, and cover with stock. Cook at 350° F (Gas Mark 4) for 45 minutes. Cool.
Pack into waxed or rigid plastic container.
To serve reheat in double boiler, or at 350°F (Gas Mark 4) for 1 hour. The gravy may be thickened a little after reheating.
Storage Time 1 month.

BARBECUED BEEF OR LAMB

2 lbs. cold meat (beef or lamb)
1 large onion
4 tablespoons butter
8 oz. tomato sauce
4 tablespoons bottled sauce
2 tablespoons brown sugar
2 tablespoons vinegar
2 teaspoons Worcestershire sauce
1 teaspoon salt
½ teaspoon celery salt
Pinch of Cayenne pepper
Dash of Tabasco sauce

Chop or sliced meat. Slice onion and cook until beginning to brown in butter. Add all ingredients except meat, and simmer for 15 minutes. Stir in meat and heat through. Cool.
Pack in waxed or rigid plastic containers.
To serve thaw at room temperature for 1 hour, then heat in double boiler or moderate oven.
Storage Time 1 month.

CREOLE LAMB

2 lbs. cold lamb
2 onions
1 garlic clove
3 dessertspoons olive oil
2 dessertspoons chili sauce
1 dessertspoon Worcestershire sauce
1 dessertspoon vinegar
Salt and pepper
Bayleaf and thyme
½ pint stock

Slice the cold lamb. Mince the onions and garlic and mix with all ingredients, except meat. Simmer for 15 minutes, then add meat and cook for 5 minutes. Cool.
Pack in waxed or rigid plastic container.
To serve thaw at room temperature for 1 hour, then heat in double boiler or moderate oven.
Storage Time 1 month.

BEEF IN CURRANT SAUCE

1 lb. cold beef	1 tablespoon suger
4 oz. redcurrant jelly	1 teaspoon dry mustard
1 oz. raisins	Salt and pepper
1 tablespoon vinegar	1 tablespoon cornflour

Slice the beef. In a double boiler, melt the redcurrant jelly and gradually work in raisins, vinegar, sugar, mustard, salt and pepper. Thicken with cornflour mixed with a little water, and cook for 10 minutes. Add beef and cook for 5 minutes. Cool.
Pack in waxed or rigid plastic container.
To serve thaw at room temperature for 1 hour, then reheat in double boiler.
Storage Time 1 month

CHICKEN IN TOMATO SAUCE

2 lbs. cooked chicken meat	½ teaspoon marjoram
1 lb. can tomatoes	1 tablespoon tomato purée
1 garlic clove	Salt and pepper
1 medium onion	1 tablespoon white wine
1 green pepper	6 drops Tabasco sauce
1 teaspoon basil	4 tablespoons olive oil

Cut chicken in neat slices. Put tomatoes through a sieve. Crush garlic and chop onion and pepper finely. Cook garlic, onion and pepper in hot olive oil until just soft. Stir in tomatoes and herbs, tomato puree, salt and pepper, white wine and Tabasco sauce. Simmer for 15 minutes. Stir in chicken and cook for 5 minutes.
Pack in waxed or rigid plastic container.
To serve thaw at room temperature for 1 hour, then reheat in double boiler.
Storage Time 1 month.

DUCK IN RED WINE

1 lb. cooked duck meat	2 tablespoons stuffed olives
2 tablespoons olive oil	½ pint stock
1 small onion	½ pint red wine
4 oz. mushrooms	¼ teaspoon thyme
1 stick celery	2 tablespoons cornflour

Heat oil and cook sliced onion, mushrooms and celery until just soft. Add sliced olives, stock, wine and thyme and simmer for 10 minutes. Add sliced duck and cook for 5 minutes, then thicken sauce with cornflour mixed with a little water. Season to taste with salt and pepper. Cool.
Pack in waxed or rigid plastic container.
To serve thaw at room temperature for 1 hour, then reheat in double boiler.
Storage Time 1 month.

POTTED CHICKEN

Cold roast chicken   Salt and pepper

Strip chicken from bones and simmer bones in a little water to make strong stock. Mince chicken finely and moisten with stock.
Pack into small foil containers, cover with foil.
To serve thaw at room temperature for 1 hour and use for toast or sandwiches, or cut in slices to serve with salad. Use immediately after thawing.
Storage Time 1 month.

VEGETABLE FLAN

8 oz. short pastry   8 oz. cooked mixed vegetables
2 oz. grated cheese   ½ pint cheese sauce

Work grated cheese into pastry, and line flan ring or foil pie dish. Mix vegetables with enough cheese sauce to bind them, and put into pastry case. Top with remaining cheese sauce. Bake at 375°F (Gas Mark 5) for 30 minutes.
Pack by covering case with foil, or by putting case into polythene bag or box to avoid crushing.
To serve put frozen flan into oven at 375°F (Gas Mark 5) and heat for 45 minutes. The flan is improved by a topping of grated cheese put on before it is reheated.
Storage Time 1 month.

INDEX

Apple and raspberry pie filling, 52, 53
  baked, 93, 94
  dumplings, baked, 47
  ice cream, 65
  juice, 123
  sauce, 110
  Swedish cake, 47
Apples, spiced, 126
Apricot jam, 125

Bacon loaf, 99
  pasties, 100
  wraps, 114
Bakewell tart, 53, 54
Baps, 80
Batch cooking, 97
Beef and sausage roll, 99, 100
  barbecued, 139
  corned, envelopes, 100, 101
  galantine, 98, 99
  in currant sauce, 140
  in wine, 34
  jellied, 34
Biscuits, 67–76
  basic sugar, 75, 76
Blackberry jam, 125
Blackcurrant flan, 53
Brandy butter, 110, 111
Bread, 79–82
  sauce, 110
Bridge rolls, 80
Brioche, 81, 82
Brownies, 74, 75

Cabbage, stuffed rolls, 138, 139
Cakes, 67–76
Canapes, 113
Casseroles, 31–33
Cheese and crab dip, 116, 117
  and crab rolls, 114, 115
  blue, spread, 116
  cake, baked, 46
  cigarettes, 114
  cream, and liver spread, 116
  cream, balls, 115
  orange dip, 117
  toasts, 115

Chestnut stuffing, 109
Chicken in cream sauce, 119
  in curry sauce, 38
  in mushroom sauce, 132
  in tomato sauce, 140
  liver paté, 27
  oven-fried, 38
  pie, 50
  positano, 133
  potted, 141
  Tetrazzini, 57, 58
  see also Coq au vin
Chocolate cake, 73
  crumb cake, 75
  mousse, 43
  pudding ice, 65
Choux pastry, 74
Christmas pudding, 108
Chutneys, 122
Citrus fruit juice, 123, 124
Cod steaks with curry sauce, 134
Cod's roe paste, 29
Coffee cake, 73
  pudding, 46
Coq au vin, 105
Coquille St. Jacques, 104, 105
Corned beef envelopes, 100, 101
Cornish pasties, 52
Cottage pie, 92
Crab and cheese dip, 116, 117
  and cheese rolls, 114, 115
Cranberry orange relish, 126
  sauce, 110
Cream cheese, *see* Cheese, cream
Cream ice, 63, 64
Croissants, 81
Cumberland tart, 54
Curry sauce, 56, 57
Custard ice, 63

Danish cherry tart, 54
  pastries, 83, 84
Date and rum ice cream, 65
Dips, 113, 114, 116, 117
Doughnuts, 83
Drop scones, 84, 85
Duck in red wine, 140, 141
Dundee cake, 71

Fish cakes, 90
  pie, 91, 119
  pie Russian, 134
  turnovers, 52
Flans, blackcurrant, 53
  savoury, 23
Freezer fudge, 95
  plan for dinner parties, 106
Fresh fruit ice, 64
Frozen foods, commercial, 130–135
Fruit cakes, 69
  cream, 44
  crumbles, 94
  fillings, 49
  fritters, 94
  juices, 122, 123
  mousse, 44
  sweets, 42
  syrups, 121, 122
  whip, 120

Galantines, 32, 33
Garnishes, 112, 113
Gelatine ice, 63
  sweets, 42
Giblets, 107, 108
Gingerbread, 73
Gnocchi, 59
Golden lemon cake, 74
Grapes, cocktail, 115, 116
Griddlebread, 85
Griddlecakes, 79
  Welsh, 85

Ham and horseradish pie, 117
Hare, jugged, 40
  paté, 28
Honey loaf, 102
Horseradish, and ham, pie, 117

Ice cream, 60–66
  bombes, 62, 63
  layer cake, 66
Ice cubes, 112
Icebox cake, 45

Jam, 122, 124, 125

Kedgeree, 91
Kidney soup, 25
Kidneys in wine, 37
Kipper paté, 133

Lamb, barbecued, 139
  creole, 139
  curry, 35
Leftovers, 97, 136–141
Lemon cake, golden, 74
  ice pie, 66
  puddings, 44
  sorbet, 64
Liver and cream cheese spread, 116
  casserole, 37
Luncheon cake, 71, 72

Macaroni, 55
  cheese, 57
Marrow, stuffed rings, 92, 93
Meat, 31–38
  balls, 37, 38
  loaf, 32, 33, 91, 92
  pie fillings, 49, 50
Mincepies, 108
Mixtures, puddings and cakes, 41, 42
Mousses, 42
Muffins, 82

Nesselrode mousse, 43

Onion and sausage pie, 100
  soup, 26
Orange and cheese dip, 117
  castles, 120
  crumb pie, 135
  honeycomb, 134
  loaf, 72
  sorbet, 64, 135
Overcooking, 20
Oxtail soup, 25

Packaging, 13, 33
  see also individual recipes
Pancakes, 79, 85, 86
  savoury, 23, 30
Pastas, 55–59
Pastry, 48, 113
  choux, 74
Patés, 23, 27–29
Peach jam, 125
  sauce, 125, 126
Peaches in white wine, 45
  stuffed, 106
Pears in red wine, 45
Pheasant in cider, 39, 40
  paté, 27
Pickles, 122

Picnic tea loaf, 101
Pies, 48–51
  hotwater crust, 50
Pigeon casserole, 39
  pie, 50
Pigeons, pot roast, 39
Plaice, mushroom stuffed, 133
Pork paté, 28
  with orange sauce, 36
Potatoes, baked stuffed, 93
  creamed, 120
  duchesse, 105, 106
Poultry, 31, 38–40, 107
  stuffing, 107, 108, 109
Preparation of cooked foods, 20, 33

Quiche Lorraine, 29

Raisin shortcake, 101
Raspberry and apple pie filling, 52, 53
  jam, 124
  sauce, 125
Records, 18
Rice, 55
  Spanish, 58, 59
Risotto, 58, 59
Rose hip syrup, 123
Rum and date ice cream, 65

Sandwiches, 87–89, 113
Sauces for pastas, 55
Sausage and beef roll, 99, 100
  and onion pie, 100
  stuffing, 109
Scallops, 104, 105
Scones, 79, 84
Scotch broth, 25
Selection for freezing, 19
Shrimp bisque, 26
Shrimps, potted, 29
Soups, 22–27

Spaghetti, 55
  sauce, 56
Spanish rice, 58, 59
Spinach, creamed, 119, 120
Sponge cakes, 68, 69, 70
  drops, 70, 71
Spreads, 113, 114
Steak and kidney pie, 50
Storage times, 11
  see also individual recipes
Storing, 18
Strawberry jam, 124
  mousse, 42, 43
  water ice, 65
Stuffing, 108, 109
Sugar cake, 72
Swedish applecake, 47
Sweets, 41–47
Syrups, 121, 122

Tea cakes, 82, 83
Temperatures for freezing, 15
Thawing, 21
  of poultry, 108
  see also individual recipes
Tomato juice, 124
  sauce, 56
  soup, 24
Turkey, creamed, 111
  roll, 111

Veal in tomato sauce, 35
  with cheese, 36
Vegetable flan, 141
  soup, 24
Vegetables, commercially frozen, 131, 132
Victoria sandwich, 70

Warnings, 18–20

Yeast breads and mixtures, 77, 78